Spectral Methods of Automorphic Forms
Second Edition

GRADUATE STUDIES
IN MATHEMATICS **53**

Spectral Methods of Automorphic Forms

Second Edition

Henryk Iwaniec

American Mathematical Society
Revista Matemática Iberoamericana

2000 *Mathematics Subject Classification*. Primary 11F12, 11F30, 11F72.

Library of Congress Cataloging-in-Publication Data

Iwaniec, Henryk.
   Spectral methods of automorphic forms / Henryk Iwaniec.—2nd ed.
      p. cm. — (Graduate studies in mathematics, ISSN 1065-7339 ; v. 53)
   First ed. published in Revista matemática iberoamericana in 1995.
   Includes bibliographical references and index.
   ISBN 0-8218-3160-7 (acid-free paper) | Softcover ISBN 978-1-4704-6622-0
   1. Automorphic functions. 2. Automorphic forms. 3. Spectral theory (Mathematics)
I. Title. II. Series.

QA353.A9 I88   2002
511.3′3—dc21   2002027749

Contents

Preface

I was captivated by a group of enthusiastic Spanish mathematicians whose desire for cultivating modern number theory I enjoyed recently during two memorable events, the first at the summer school in Santander, 1992, and the second while visiting the Universidad Autónoma in Madrid in June 1993. These notes are an expanded version of a series of eleven lectures I delivered in Madrid[1]. They are more than a survey of favorite topics, since proofs are given for all important results. However, there is a lot of basic material which should have been included for completeness, but was not because of time and space limitations. Instead, to make a comprehensive exposition we focus on issues closely related to the the spectral aspects of automorphic forms (as opposed to the arithmetical aspects, to which I intend to return on another occasion).

Primarily, the lectures are addressed to advanced graduate students. I hope the student will get inspiration for his own adventures in the field. This is a goal which Professor Antonio Córdoba has a vision of pursuing throughout the new volumes to be published by the Revista Matemática Iberoamericana. I am pleased to contribute to part of his plan.

Many people helped me prepare these notes for publication. In particular I am grateful to Fernando Chamizo, José Luis Fernández, Charles Mozzochi, Antonio Sánchez-Calle and Nigel Pitt for reading and correcting an early draft. I also acknowledge the substantial work in the technical preparation of this text by Domingo Pestana and María Victoria Melián, without which this book would not exist.

New Brunswick, October 1994 Henryk Iwaniec

---

[1] I would like to thank the participants and the Mathematics Department for their warm hospitality and support.

Preface to the AMS Edition

Since the initial publication by the Revista Matemática Iberoamericana I have used this book in my graduate courses at Rutgers several times. I have heard from my colleagues that they also had used it in their teaching. On several occasions I have been told that the book would be a good source of ideas, results and references for graduate students, if it were easier to obtain. The American Mathematical Society suggested publishing an edition, which would increase its availability. Sergei Gelfand should be given special credit for his persistence during the past five years to convince me of the need for a second publication. I am also very indebted to the editors of the Revista for the original edition and its distribution and for the release of the copyright to me, so this revised edition could be realized.

I eliminated all the misprints and mistakes which were kindly called to my attention by several people. I also changed some words and slightly expanded some arguments for better clarity. In a few places I mentioned recent results. I did not try to include all of the best achievements, since this had not been my intention in the first edition either. Such a task would require several volumes, and that could discourage beginners. I hope that a subject as great as the analytic theory of automorphic forms will eventually be presented more substantially, perhaps by several authors. Until that happens, I am recommending that newcomers read the recent survey articles and new books which I have listed at the end of the regular references.

New Brunswick, June 2002   Henryk Iwaniec

Introduction

The concept of an automorphic function is a natural generalization of that of a periodic function. Furthermore, an automorphic form is a generalization of the exponential function

$$e(z) = e^{2\pi i z}.$$

To define an automorphic function in an abstract setting, one needs a group Γ acting discontinuously on a locally compact space X; the functions on X which are invariant under the group action are called automorphic functions (the name was given by F. Klein in 1890). A typical case is the homogeneous space $X = G/K$ of a Lie group G, where K is a closed subgroup. In this case the differential calculus is available, since X is a riemannian manifold. The automorphic functions which are eigenfunctions of all invariant differential operators (these include the Laplace operator) are called automorphic forms. The main goal of harmonic analysis on the quotient space $\Gamma \backslash X$ is to decompose every automorphic function satisfying suitable growth conditions into automorphic forms. In these lectures we shall present the basic theory for Fuchsian groups acting on the hyperbolic plane.

When the group Γ is arithmetic, there are interesting consequences for number theory. What makes a group arithmetic is the existence of a large family (commutative algebra) of certain invariant, self-adjoint operators, the Hecke operators. We shall get into this territory only briefly in Sections 8.4 and 13.3 to demonstrate its tremendous potential. Many important topics lie beyond the scope of these lectures; for instance, the theory of automorphic L-functions is omitted entirely.

A few traditional applications are included without straining for the best results. For more recent applications the reader is advised to see the original sources (see the surveys [Iw 1, 2] and the book [Sa 3]).

1

There are various books on spectral aspects of automorphic functions, but none covers and treats in detail as much as the expansive volumes by Dennis Hejhal [He1]. I recommend them to anyone who is concerned with research. In these books one also finds a very comprehensive bibliography. Those who wish to learn about the theory of automorphic forms on other symmetric spaces in addition to the hyperbolic plane should read Audrey Terras [Te]. A broad survey with emphasis on new developments is given by A. B. Venkov [Ve].

Harmonic Analysis on the Euclidean Plane

We begin by presenting the familiar case of the euclidean plane

$$\mathbb{R}^2 = \left\{ (x, y) : \ x, y \in \mathbb{R} \right\}.$$

The group $G = \mathbb{R}^2$ acts on itself as translations, and it makes $\mathbb{R}^2$ a homogeneous space. The euclidean plane carries the metric

$$ds^2 = dx^2 + dy^2$$

of curvature $K = 0$, and the Laplace-Beltrami operator associated with this metric is given by

$$D = \frac{\partial^2}{\partial x^2} + \frac{\partial^2}{\partial y^2}.$$

Clearly the exponential functions

$$\varphi(x, y) = e(ux + vy), \qquad (u, v) \in \mathbb{R}^2,$$

are eigenfunctions of D;

$$(D + \lambda)\varphi = 0, \qquad \lambda = \lambda(\varphi) = 4\pi^2(u^2 + v^2).$$

The well-known Fourier inversion

$$\hat{f}(u, v) = \iint f(x, y) \, e(ux + vy) \, dx \, dy,$$

$$f(x, y) = \iint \hat{f}(u, v) \, e(-ux - vy) \, du \, dv,$$

is just the spectral resolution of D on functions satisfying proper decay conditions.

Another view of this matter is offered by invariant integral operators

$$(Lf)(z) = \int_{\mathbb{R}^2} k(z, w)\, f(w)\, dw.$$

For L to be G-invariant it is necessary and sufficient that the kernel function, $k(z, w)$, depends only on the difference $z - w$, i.e. $k(z, w) = k(z - w)$. Such an L acts by convolution: $Lf = k * f$. One shows that the invariant integral operators mutually commute and that they commute with the Laplace operator as well. Therefore the spectral resolution of D can be derived from that for a sufficiently large family of invariant integral operators. By direct computation one shows that the exponential function $\varphi(x, y) = e(ux + vy)$ is an eigenfunction of L with eigenvalue $\lambda(\varphi) = \hat{k}(u, v)$, the Fourier transform of $k(z)$.

Of particular interest will be the radially symmetric kernels:

$$k(x, y) = k(x^2 + y^2), \qquad k(r) \in C_0^\infty(\mathbb{R}^+).$$

Using polar coordinates, one finds that the Fourier transform is also radially symmetric. More precisely,

$$\hat{k}(u, v) = \pi \int_0^{+\infty} k(r)\, J_0(\sqrt{\lambda r})\, dr,$$

where $\lambda = 4\pi^2(u^2 + v^2)$ and $J_0(z)$ is the Bessel function

$$J_0(z) = \frac{1}{\pi} \int_0^\pi \cos(z \cos \alpha)\, d\alpha.$$

Classical analytic number theory benefits a lot from harmonic analysis on the torus $\mathbb{Z}^2 \backslash \mathbb{R}^2$ (which is derived from that on the free space $\mathbb{R}^2$ by the unfolding technique), as it exploits properties of periodic functions. Restricting the domain of the invariant integral operator L to periodic functions, we can write

$$(Lf)(z) = \int_{\mathbb{Z}^2 \backslash \mathbb{R}^2} K(z, w)\, f(w)\, dw,$$

where

$$K(z, w) = \sum_{p \in \mathbb{Z}^2} k(z + p, w),$$

by folding the integral. Hence the trace of L on the torus is equal to

$$\text{Trace}\, L = \int_{\mathbb{Z}^2 \backslash \mathbb{R}^2} K(w, w)\, dw = \sum_{p \in \mathbb{Z}^2} k(p) = \sum_{m, n \in \mathbb{Z}} k(m, n).$$

On the other hand, by the spectral decomposition (classical Fourier series expansion)

$$K(z, w) = \sum_{\varphi} \lambda(\varphi) \, \varphi(z) \, \overline{\varphi}(w);$$

the trace is given by

$$\text{Trace } L = \sum_{\varphi} \lambda(\varphi) = \sum_{u,v \in \mathbb{Z}} \hat{k}(u, v).$$

Comparing both results, we get the trace formula

$$\sum_{m,n \in \mathbb{Z}} k(m, n) = \sum_{u,v \in \mathbb{Z}} \hat{k}(u, v),$$

which is better known as the Poisson summation formula. By a linear change of variables this formula can be modified for sums over general lattices $\Lambda \subset \mathbb{R}^2$. On both sides of the trace formula on the torus $\Lambda \backslash \mathbb{R}^2$ the terms are of the same type because the geometric and the spectral points range over dual lattices. However, one loses the self-duality on negatively curved surfaces, although the relevant trace formula is no less elegant (see Theorem 10.2).

In particular, for a radially symmetric function the Poisson summation becomes

Theorem (Hardy-Landau, Voronoi). *If $k \in C_0^\infty(\mathbb{R})$, then*

$$\sum_{\ell=0}^{\infty} r(\ell) \, k(\ell) = \sum_{\ell=0}^{\infty} r(\ell) \, \tilde{k}(\ell),$$

where $r(\ell)$ denotes the number of ways to write ℓ as the sum of two squares,

$$r(\ell) = \#\{(m, n) \in \mathbb{Z}^2 : m^2 + n^2 = \ell\},$$

and $\tilde{k}$ is the Hankel type transform of k given by

$$\tilde{k}(\ell) = \pi \int_0^{+\infty} k(t) \, J_0(2\pi \sqrt{\ell t}) \, dt.$$

Note that the lowest eigenvalue $\lambda(1) = 4\pi^2 \ell$ with $\ell = 0$ for the constant eigenfunction $\varphi = 1$ contributes

$$\tilde{k}(0) = \pi \int_0^{+\infty} k(t) \, dt,$$

which usually constitutes the main term. Taking a suitable kernel (a smooth approximation to a step function) and using standard estimates for Bessel's function, we derive the formula

$$\sum_{\ell \leq x} r(\ell) = \pi x + O(x^{1/3}),$$

which was originally established by Voronoi and Sierpinski by different means. The left side counts integral points in the circle of radius $\sqrt{x}$. This is also equal to the number of eigenvalues $\lambda(\varphi) \leq 4\pi^2 x$ (counted with multiplicity), so we have

$$\#\{\varphi : \lambda(\varphi) \leq T\} = \frac{T}{4\pi} + O(T^{1/3}).$$

In view of the above connection the Gauss circle problem becomes the Weyl law for the operator D (see Section 11.1).

Chapter 1

Harmonic Analysis on the Hyperbolic Plane

1.1. The upper half-plane

As a model of the hyperbolic plane we will use the upper half of the plane $\mathbb{C}$ of complex numbers:

$$\mathbb{H} = \left\{ z = x + iy : \ x \in \mathbb{R}, \ y \in \mathbb{R}^+ \right\}.$$

$\mathbb{H}$ is a riemannian manifold with the metric derived from the Poincaré differential,

$$(1.1) \qquad\qquad ds^2 = y^{-2}(dx^2 + dy^2).$$

The distance function on $\mathbb{H}$ is given explicitly by

$$(1.2) \qquad\qquad \rho(z, w) = \log \frac{|z - \overline{w}| + |z - w|}{|z - \overline{w}| - |z - w|}.$$

We have

$$(1.3) \qquad\qquad \cosh \rho(z, w) = 1 + 2\, u(z, w),$$

where

$$(1.4) \qquad\qquad u(z, w) = \frac{|z - w|^2}{4 \operatorname{Im} z \operatorname{Im} w}.$$

This function (a point-pair invariant) is more practical than the true distance function $\rho(z, w)$.

To describe the geometry of $\mathbb{H}$ we shall use well-known properties of the Möbius transformations

$$(1.5) \qquad gz = \frac{az + b}{cz + d}, \qquad a, b, c, d \in \mathbb{R}, \ ad - bc = 1.$$

Observe that a Möbius transformation g determines the matrix $\begin{pmatrix} a & b \\ c & d \end{pmatrix}$ up to sign. In particular, both matrices $1 = \begin{pmatrix} 1 & \\ & 1 \end{pmatrix}$ and $-1 = \begin{pmatrix} -1 & \\ & -1 \end{pmatrix}$ give the identity transformation. We shall always take this distinction into account, but often without mention.

Throughout we denote $G = SL_2(\mathbb{R})$, the group of real matrices of determinant 1. The group $PSL_2(\mathbb{R}) = G/(\pm 1)$ of all Möbius transformations acts on the whole compactified complex plane $\hat{\mathbb{C}} = \mathbb{C} \cup \{\infty\}$ (the Riemann sphere) as conformal mappings. A Möbius transformation g maps a euclidean circle onto a circle subject to the convention that the euclidean lines in $\hat{\mathbb{C}}$ are also circles. Of course, the centers may not be preserved, since g is not a euclidean isometry, save for $g = \pm \begin{pmatrix} 1 & * \\ & 1 \end{pmatrix}$, which represents a translation.

If $g = \begin{pmatrix} * & * \\ c & d \end{pmatrix} \in G$, then

$$(1.6) \qquad gz - gw = \frac{z - w}{(cz + d)(cw + d)}.$$

In particular, this shows that

$$|gz - gw| = |z - w|,$$

if both points are on the curve

$$(1.7) \qquad C_g = \left\{ z \in \mathbb{C} : \ |cz + d| = 1 \right\}.$$

If $c \neq 0$, this is a circle centered at $-d/c$ of radius $|c|^{-1}$. Hence C_g is the locus of points z such that the line element at z is not altered in euclidean length by the motion g; therefore g acts on C_g as a euclidean isometry. Naturally C_g is called the *isometric circle* of g. By (1.6) we get

$$(1.8) \qquad \frac{d}{dz} gz = (cz + d)^{-2},$$

so we call $|cz + d|^{-2}$ the deformation of g at z. In this language the interior of C_g consists of points with deformation greater than 1 and the exterior

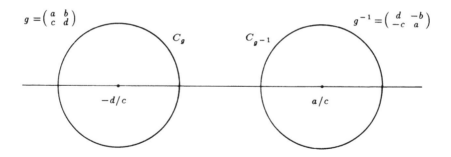

Figure 1. Isometric circles.

consists of points with deformation less than 1. Note that g maps C_g to $C_{g^{-1}}$ and reverses the interior of C_g onto the exterior of $C_{g^{-1}}$.

For $g = \left(\begin{smallmatrix} * & * \\ c*d \end{smallmatrix} \right) \in G$ we introduce the function

$$(1.9) \qquad j_g(z) = cz + d.$$

The j-function satisfies the chain rule of differentiation,

$$(1.10) \qquad j_{gh}(z) = j_g(hz)\, j_h(z).$$

It follows from the formula

$$(1.11) \qquad |j_g(z)|^2 \operatorname{Im} gz = \operatorname{Im} z$$

that the complex plane $\hat{\mathbb{C}}$ splits into three G-invariant subspaces, name- ly $\mathbb{H}$, $\overline{\mathbb{H}}$ (the lower half-plane) and $\hat{\mathbb{R}} = \mathbb{R} \cup \{\infty\}$ (the real line, the common boundary of $\mathbb{H}$ and $\overline{\mathbb{H}}$). Moreover, we have

$$(\operatorname{Im} gz)^{-1}|dgz| = (\operatorname{Im} z)^{-1}|dz|\,,$$

which shows that the differential form (1.1) on $\mathbb{H}$ is G-invariant. This implies that the Möbius transformations are isometries of the hyperbolic plane. In addition to these isometries we have the reflection in the imaginary line, $z \mapsto -\bar{z}$, which reverses the orientation. Using the above properties, one can prove

Theorem 1.1. *The whole group of isometries of $\mathbb{H}$ is generated by the Möbius transformations and the reflection $z \mapsto -\bar{z}$.*

Theorem 1.2. *The hyperbolic lines (geodesics in $\mathbb{H}$) are represented by the euclidean semi-circles and half-lines orthogonal to $\mathbb{R}$.*

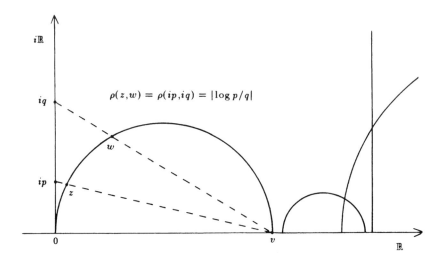

Figure 2.. Geodesics in $\mathbb{H}$.

The hyperbolic circles (loci of points at a fixed distance from a given point in $\mathbb{H}$) are represented by the euclidean circles in $\mathbb{H}$ (of course, not with the same centers).

There are various interesting relations in the hyperbolic plane. For instance, the trigonometry for a triangle asserts that

$$\frac{\sin\alpha}{\sinh a} = \frac{\sin\beta}{\sinh b} = \frac{\sin\gamma}{\sinh c}$$

and

$$\sin\alpha \, \sin\beta \, \cosh c = \cos\alpha \, \cos\beta + \cos\gamma \,,$$

where α, β, γ are the interior angles from which the sides of length a, b, c are seen, respectively. The latter relation reveals that the length of sides depends only on the interior angles.

More counter-intuitive features occur with the area. To define area, one needs a measure. The riemannian measure derived from the Poincaré differential $ds = y^{-1}|dz|$ on $\mathbb{H}$ is expressed in terms of the Lebesgue measure simply by

(1.12) $d\mu z = y^{-2}\, dx\, dy.$

It is easy to show directly by (1.8) and (1.11) that the above measure is G-invariant.

Theorem 1.3 (Gauss defect). *The area of a hyperbolic triangle with interior angles α, β, γ is equal to*

(1.13) $\pi - \alpha - \beta - \gamma \,.$

There is a universal inequality between the area and the boundary length of a domain in a riemannian surface called the isoperimetric inequality; it asserts that

$$4\pi A - K A^2 \le L^2,$$

where A is the area, L is the length of the boundary, and K is the curvature (assumed to be constant). The isoperimetric inequality is sharp, since the equality is attained for discs. In the euclidean plane we have $K = 0$ and $4\pi A \le L^2$, so the area can be much larger than the length of the boundary. On the other hand, in the hyperbolic plane we have $K = -1$ and

(1.14) $$4\pi A + A^2 \le L^2.$$

Hence $A \le L$, so the hyperbolic area is comparable to the boundary length. This observation should explain why the analysis on $\mathbb{H}$ is more subtle than that on $\mathbb{R}^2$ in various aspects. For example, the lattice point problem on $\mathbb{H}$ is much harder than that on $\mathbb{R}^2$ (see Chapter 12).

To illustrate the rules, look at the hyperbolic disc of radius r centered at i (the origin of $\mathbb{H}$, so to speak).

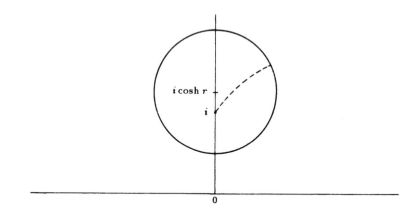

Figure 3. A hyperbolic disc.

The hyperbolic area is $4\pi(\sinh(r/2))^2$, and the circumference is $2\pi \sinh r$. On the other hand, the euclidean center is at $i \cosh r$ and the radius is $\sinh r$, so the area is $\pi(\sinh r)^2$ and the circumference is $2\pi \sinh r$. Although the circumferences are the same, the euclidean area is much larger than the hyperbolic area if r is large (still approximately the same area for small r). Most of the hyperbolic area concentrates along a lower segment of the boundary.

1.2. $\mathbb{H}$ as a homogeneous space

Occasionally it will be convenient to work with the homogeneous space model of the hyperbolic plane rather than the Poincaré upper half-plane. Here we describe that model.

The group $G = SL_2(\mathbb{R})$ acts on $\mathbb{H}$ transitively, so $\mathbb{H}$ is obtained as the orbit of a point:

$$\mathbb{H} = Gz = \{gz : \ g \in G\}.$$

The point $z = i$ is special; its stability group is the orthogonal group

$$K = \{k \in G : \ ki = i\} = \left\{k(\theta) = \begin{pmatrix} \cos\theta & \sin\theta \\ -\sin\theta & \cos\theta \end{pmatrix} : \ \theta \in \mathbb{R}\right\}.$$

The element $k(\theta) \in K$ acts on $\mathbb{H}$ as the rotation at i of angle 2θ.

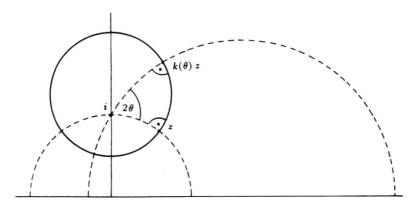

Figure 4. The action of K.

The upper half-plane $\mathbb{H}$ can be identified with the quotient G/K (the space of orbits) so that a point $z \in \mathbb{H}$ corresponds to the coset gK of all motions which send i to z. In such a realization of $\mathbb{H}$ the group G acts on itself by matrix multiplication.

In order to be able to use both models alternatively and consistently, we need an explicit connection between the rectangular coordinates of points in $\mathbb{H}$ and the matrix entries of group elements in G. This is given through the Iwasawa decomposition

$$G = N\,A\,K.$$

Here A and N are the following subgroups:

$$A = \left\{ \begin{pmatrix} a & \\ & a^{-1} \end{pmatrix} : \ a \in \mathbb{R}^+ \right\},$$

$$N = \left\{ \begin{pmatrix} 1 & x \\ & 1 \end{pmatrix} : x \in \mathbb{R} \right\}.$$

The Iwasawa decomposition asserts that any $g \in G$ has the unique factorization

$$g = n\,a\,k, \qquad n \in N,\ a \in A,\ k \in K.$$

To see this, first use a and k to make a matrix with a given lower row,

$$\begin{pmatrix} a & * \\ & a^{-1} \end{pmatrix} \begin{pmatrix} * & * \\ -\sin\theta & \cos\theta \end{pmatrix} = \begin{pmatrix} * & * \\ \gamma & \delta \end{pmatrix},$$

i.e. take $a = (\gamma^2 + \delta^2)^{-1/2}$ and θ such that $-\sin\theta = \gamma\,a$, $\cos\theta = \delta a$. Then apply on the left a suitable translation $\begin{pmatrix} 1 & x \\ & 1 \end{pmatrix}$ to arrive at the desired upper row without altering the lower row. Since the above procedure is unique at each step, this proves the Iwasawa decomposition.

It is clear that a point $z = x + iy$ in $\mathbb{H}$ corresponds to the coset gK in G/K for $g = n\,a\,k$ with

$$n = n(x) = \begin{pmatrix} 1 & x \\ & 1 \end{pmatrix}, \qquad a = a(y) = \begin{pmatrix} y^{1/2} & \\ & y^{-1/2} \end{pmatrix}.$$

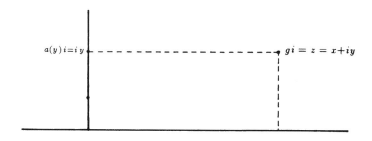

Figure 5. The Iwasawa coordinates.

The group A operates on $\mathbb{H}$ by dilations, and the group N operates by translations. The upper half-plane is identified through the Iwasawa decomposition with the group of upper triangular matrices of determinant 1:

$$P = \left\{ \begin{pmatrix} * & * \\ & * \end{pmatrix} \in G \right\}.$$

We have $\mathbb{H} = G/K = N\,A = A\,N = P$. Notice that A and N are abelian, whereas P is not, yet the following commutativity relation holds:

$$a(y)\,n(x) = n(xy)\,a(y).$$

To perform integration on P, one needs a measure. We proceed by recalling a few facts about topological groups.

Suppose G is a locally compact group. Then G has a left-invariant measure dg, say, which means that

$$\int f(hg)\, dg = \int f(g)\, dg$$

for any test function integrable on G. The left-invariant measure is unique up to a positive constant multiple; therefore,

$$\int f(gh)\, dg = \delta(h) \int f(g)\, dg\,,$$

where $\delta(h) > 0$ depends only on h because dgh^{-1} is another left-invariant measure. The factor $\delta(h)$ is called *the modular function* of G. Clearly $\delta : G \longrightarrow \mathbb{R}^+$ is a group homomorphism, and one also shows that δ is continuous. Similarly, G has a right-invariant measure; it is equal to $\delta(g)\, dg$ up to a constant factor. If $\delta(g) = 1$, then G is called *unimodular*. Abelian groups are obviously unimodular, and compact groups are also unimodular because the multiplicative group $\mathbb{R}^+$ contains no compact subgroups other than the trivial one.

Now we return to the group $G = SL_2(\mathbb{R})$ and its decomposition $G = NAK$. Each factor, being abelian, is a unimodular group. The invariant measures on N, A, K are given by

$$dn = dx \qquad\qquad \text{if}\ \ n = n(x) = \begin{pmatrix} 1 & x \\ & 1 \end{pmatrix},$$

$$da = y^{-1}\, dy \qquad\qquad \text{if}\ \ a = a(y) = \begin{pmatrix} \sqrt{y} & \\ & 1/\sqrt{y} \end{pmatrix},$$

$$dk = (2\pi)^{-1} d\theta \qquad \text{if}\ \ k = k(\theta) = \begin{pmatrix} \cos\theta & \sin\theta \\ -\sin\theta & \cos\theta \end{pmatrix},$$

where dx, dy, $d\theta$ are the Lebesgue measures. Since K is compact, we could normalize the measure on K to have $\int_K dk = 1$.

Let us define a measure dp on $P = AN$ by requiring that

$$\int_P f(p)\, dp = \int_A \int_N f(an)\, da\, dn\,;$$

i.e. $dp = y^{-1} dx\, dy$ in rectangular coordinates. We shall show that dp is left-invariant. First we need a multiplication rule on $P = AN$. Given $h = a(u)\, n(v)$ and $p = a(y)\, n(x)$, one has $hp = a(uy)\, n(x + vy^{-1})$. Hence,

$$\int_P f(hp)\, dp = \int_A \int_N f(a(uy)\, n(x + vy^{-1})) \frac{dx\, dy}{y}$$

$$= \int_A \int_N f(a(y)\,n(x))\,\frac{dx\,dy}{y} = \int_P f(p)\,dp\,,$$

which shows that dp is left-invariant.

Furthermore by Fubini's theorem we derive the relation

$$\int_A \int_N f(a(y)\,n(x))\,\frac{dx\,dy}{y} = \int_N \int_A f(n(xy)\,a(y))\,\frac{dx\,dy}{y}$$

$$= \int_N \int_A f(n(x)\,a(y))\,\frac{dx\,dy}{y^2}\,.$$

This shows that the modular function of P is equal to $\delta(p) = y^{-1}$ if $p = a(y)\,n(x)$. Hence the right-invariant measure on P is equal to $\delta(p)\,dp = y^{-2}dx\,dy$, which is just the riemannian measure on $\mathbb{H}$.

Remark. The whole group $G = N\,A\,K = SL_2(\mathbb{R})$ is unimodular in spite of being non-abelian and not compact. One can show that the measure dg defined by

$$\int_G f(g)\,dg = \int_A \int_N \int_K f(a\,n\,k)\,da\,dn\,dk$$

is the invariant measure on G.

1.3. The geodesic polar coordinates.

We shall often encounter functions on $\mathbb{H}$ which depend only on the hyperbolic distance. Naturally, it is more convenient to work with such functions in geodesic polar coordinates rather than in rectangular coordinates $z = x + iy$. The geodesic polar coordinates are derived from Cartan's decomposition

$$G = K\,A\,K\,.$$

This asserts that any $g \in G$ can be brought to a diagonal matrix by multiplying on both sides with orthogonal matrices. To see this, first multiply on the left by $k_1 \in K$ to bring g to a symmetric matrix $g_1 = k_1\,g$. Then by conjugation in K the symmetric matrix g_1 can be brought to a diagonal matrix $a = k\,g_1\,k^{-1}$ (this is the spectral theorem for symmetric matrices). Hence, we have Cartan's decomposition $g = k_1^{-1}\,k^{-1}\,a\,k$.

We shall write any $g \in PSL_2(\mathbb{R})$ as $g = k(\varphi)\,a(e^{-r})\,k(\theta)$, with

$$k(\varphi) = \begin{pmatrix} \cos\varphi & \sin\varphi \\ -\sin\varphi & \cos\varphi \end{pmatrix}, \qquad 0 \le \varphi < \pi\,,$$

$$a(e^{-r}) = \begin{pmatrix} e^{-r/2} & \\ & e^{r/2} \end{pmatrix}, \qquad r \ge 0\,,$$

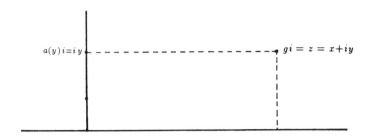

Figure 6.   The geodesic polar coordinates.

and $k(\theta)$ as in the Iwasawa decomposition. Of course, $a(e^{-r})$ is different from the one in the rectangular coordinates. We have $\rho(gi, i) = \rho(k(\varphi)\, e^{-r}i, i) = \rho(e^{-r}i, i) = r$; therefore r is the hyperbolic distance from i to $gi = z = x+iy$.

Since $k(\varphi)$ acts by rotation at i of angle 2φ, it follows that when φ ranges over $[0, \pi)$ and r over $(0, +\infty)$, the upper half-plane is covered once except for the origin $z = i$. The level curves of r and φ are orthogonal circles.

The pair (r, φ) is called the *geodesic polar coordinates* of the point z. These are related to the rectangular coordinates by $k(\varphi)\, e^{-r}i = x + iy$, *i.e.*

$$y = (\cosh r + \sinh r\, \cos 2\varphi)^{-1},$$
$$x = y\, \sinh r\, \sin 2\varphi.$$

The length element and the measure are expressed in the new coordinates as follows:

(1.15) $$ds^2 = dr^2 + (2\, \sinh r)^2\, d\varphi^2,$$
(1.16) $$d\mu z = (2\, \sinh r)\, dr\, d\varphi.$$

In the (u, φ) coordinates, where $\cosh r = 1 + 2u$ as in (1.3), we have

(1.17) $$d\mu z = 4\, du\, d\varphi.$$

1.4. Group decompositions

We have just introduced two distinct decompositions of $G = SL_2(\mathbb{R})$, namely $G = NAK$ (Iwasawa) and $G = KAK$ (Cartan). These decompositions can be generalized and unified in terms of the Lie algebra of G. Briefly, a decomposition of type $G = P_1\, Q\, P_2$, where the P_j are 1-parameter subgroups (P_j is necessarily conjugate to $\{\exp(tX_j) :\ t \in \mathbb{R}\}$ for some X_j of zero trace) and Q is a 1-dimensional subspace, yields a parametrization of G. Such a parametrization makes it possible to reduce the analysis on G to a simpler one on the components in the following fashion: put a suitable test function on the central component and use characters on the outer components.

There is yet a different type of decomposition (à la Bruhat) that we wish to consider, namely

$$G = NAN \ \cup \ N\omega AN,$$

where

$$\omega = \begin{pmatrix} & -1 \\ 1 & \end{pmatrix}$$

is the involution. Here the first part $NAN = AN = NA$ consists of the upper triangular matrices. The second part $N\omega AN$ asserts that every $g = \begin{pmatrix} a & * \\ c & d \end{pmatrix}$ with $c \neq 0$ factors uniquely into

(1.18)
$$g = \begin{pmatrix} 1 & a/c \\ & 1 \end{pmatrix} \begin{pmatrix} & -1/c \\ c & \end{pmatrix} \begin{pmatrix} 1 & d/c \\ & 1 \end{pmatrix}.$$

The three decompositions we have singled out for these lectures are destined to specific tasks. In particular, we shall employ the Iwasawa decomposition (the rectangular coordinates) for the Fourier development of automorphic forms, the Cartan decomposition (the geodesic polar coordinates) to study the Green function, and the Bruhat type decomposition (or rather its ramification—a double coset decomposition, see Section 2.4) for creating Kloosterman sums.

1.5. The classification of motions

The Möbius transformations are rigid motions of the hyperbolic plane, and they move points in distinct ways. We shall give a characterization by various means. First, notice that conjugate motions act similarly; therefore the classification should be invariant under conjugation. Given $g \in PSL_2(\mathbb{R})$, we denote its conjugacy class by

$$\{g\} = \left\{ \tau g \tau^{-1} : \ \tau \in PSL_2(\mathbb{R}) \right\}.$$

The identity motion forms a class of itself; there is nothing to examine in this class.

An important geometric invariant of the conjugation is the number of fixed points (and their configuration). Any

$$g = \begin{pmatrix} a & b \\ c & d \end{pmatrix} \neq \pm 1$$

has one or two fixed points in $\hat{\mathbb{C}}$: these fixed points are $b/(d-a)$ if $c = 0$ and $(a - d \pm \sqrt{(a+d)^2 - 4})/2c$ if $c \neq 0$. Hence there are three cases:

(i) g has one fixed point on $\hat{\mathbb{R}}$.

(ii) g has two distinct fixed points on $\hat{\mathbb{R}}$.

(iii) g has a fixed point in $\mathbb{H}$ and its complex conjugate in $\overline{\mathbb{H}}$.

Accordingly g is said to be *parabolic, hyperbolic,* or *elliptic,* and naturally this classification applies to the conjugacy classes. Every conjugacy class $\{g\}$ contains a representative in one of the groups N, A, K. The elements of $\pm N$, $\pm A$, K other that ± 1 are parabolic, hyperbolic and elliptic respectively. They act on $\mathbb{H}$ simply as follows:

(i) $z \mapsto z + v\,,$ $v \in \mathbb{R}$ (translation, fixed point ∞),

(ii) $z \mapsto p\,z\,,$ $p \in \mathbb{R}^+$ (dilation, fixed points $0, \infty$),

(iii) $z \mapsto k(\theta)\,z\,,$ $\theta \in \mathbb{R}$ (rotation, fixed point i).

A parabolic motion has infinite order, and it moves points along *horocycles* (circles in $\mathbb{H}$ tangent to $\hat{\mathbb{R}}$).

A hyperbolic motion has infinite order too; it moves points along *hypercycles* (the segments in $\mathbb{H}$ of circles in $\hat{\mathbb{C}}$ through the fixed points on $\hat{\mathbb{R}}$). The geodesic line through the fixed points of a hyperbolic motion g is mapped to itself, not identically. Of the two fixed points, one is repelling and the other is attracting. The dilation factor p is called the *norm* of g. For any z on this geodesic, $|\log p|$ is the hyperbolic distance between z and gz.

An elliptic motion may have finite or infinite order; it moves points along circles centered at its fixed point in $\mathbb{H}$.

An important algebraic invariant of conjugation is the trace, more precisely, its absolute value, because g determines the matrix $\left(\begin{smallmatrix} a & b \\ c & d \end{smallmatrix}\right)$ up to sign. In terms of trace the above classes (other than the identity) are characterized as follows:

(i) g is parabolic if and only if $|a + d| = 2\,,$

(ii) g is hyperbolic if and only if $|a + d| > 2\,,$

(iii) g is elliptic if and only if $|a + d| < 2\,.$

The rigid motions usually do not commute, but clearly if $gh = hg$, then the fixed points of g are also the fixed points of h. The converse is also true: precisely, two motions other than the identity commute if and only if they have the same set of fixed points. Hence the centralizer $\{h : gh = hg\}$ of a parabolic (hyperbolic, elliptic) motion g is just the subgroup of all the parabolic (hyperbolic, elliptic) motions which fix the same points as g, together with the identity motion (for a quick proof, consider g in one of the groups N, A, K).

1.6. The Laplace operator

Denote by T_g the following operator:

$$(T_g f)(z) = f(gz) \,.$$

A linear operator L acting on functions $f : \mathbb{H} \longrightarrow \mathbb{C}$ is said to be *invariant* if it commutes with all T_g, *i.e.*

$$L(f(gz)) = (Lf)(gz) \,, \qquad \text{for all} \ \ g \in G \,.$$

The invariant differential operators are particularly important; among these, the Laplace-Beltrami operator is special. In general, on a riemannian manifold, the Laplace-Beltrami operator Δ is characterized by the property that a diffeomorphism is an isometry if and only if it leaves Δ invariant. On the hyperbolic plane $\mathbb{H}$ the Laplace operator derived from the differential $ds^2 = y^{-2}(dx^2 + dy^2)$ is given by

$$(1.19) \qquad \Delta = y^2 \Big(\frac{\partial^2}{\partial x^2} + \frac{\partial^2}{\partial y^2} \Big) = -(z - \bar{z})^2 \frac{\partial}{\partial z} \frac{\partial}{\partial \bar{z}} \,,$$

where $\partial/\partial z = (\partial/\partial x - i\partial/\partial y)/2$ and $\partial/\partial \bar{z} = (\partial/\partial x + i\partial/\partial y)/2$ are the partial complex derivatives. Check directly that Δ is invariant.

In geodesic polar coordinates (r, φ) the Laplace operator takes the form

$$(1.20) \qquad \Delta = \frac{\partial^2}{\partial r^2} + \frac{1}{\tanh r} \frac{\partial}{\partial r} + \frac{1}{(2 \sinh r)^2} \frac{\partial^2}{\partial \varphi^2} \,.$$

Changing r into u, where $\cosh r = 2u + 1$ (see (1.3) and (1.4)), we get

$$(1.21) \qquad \Delta = u(u + 1) \frac{\partial^2}{\partial u^2} + (2u + 1) \frac{\partial}{\partial u} + \frac{1}{16 \, u(u + 1)} \frac{\partial^2}{\partial \varphi^2} \,.$$

Any differential operator on $\mathbb{H}$ which is G-invariant is a polynomial in Δ with constant coefficientes, *i.e.* the algebra of invariant differential operators is generated by Δ. A great deal of harmonic analysis on $\mathbb{H}$ concerns decomposition of functions $f : \mathbb{H} \longrightarrow \mathbb{C}$ into eigenpackets of Δ (an analogue of Fourier inversion).

1.7. Eigenfunctions of Δ

A function $f : \mathbb{H} \longrightarrow \mathbb{C}$ with continuous partial derivatives of order 2 is an eigenfunction of Δ with eigenvalue $\lambda \in \mathbb{C}$ if

$$(1.22) \qquad (\Delta + \lambda)f = 0 \,.$$

Since Δ is an elliptic operator with real-analytic coefficients, it forces its eigenfunctions to be real-analytic. The eigenfunctions with eigenvalue $\lambda = 0$ are harmonic functions; among them are holomorphic functions, *i.e.* those annihilated by the operator $\partial/\partial\bar{z}$ (the Cauchy-Riemann equations).

There are various ways of constructing eigenfunctions of Δ for a given eigenvalue λ. The standard method uses separation of variables. However, a more prolific one is the method of images; it generates a lot of eigenfunctions out of a fixed $f(z)$ by shifting to $f(gz)$, and still more by averaging $f(gz)$ over selected g in G. In this way one may search for eigenfunctions which satisfy desirable transformation rules. Either way, the result depends on the type of coordinates in which the construction is performed.

We first work in the rectangular coordinates $z = x + iy$. If one wants $f(z)$ to be a function in y only, *i.e.* constant in x, one has the obvious choice of two linearly independent solutions to (1.22), namely

$$(1.23) \qquad \frac{1}{2}(y^s + y^{1-s}) \qquad \text{and} \qquad \frac{1}{2s-1}(y^s - y^{1-s}),$$

where $s(1-s) = \lambda$. Note that $s \mapsto \lambda$ is a double cover of $\mathbb{C}$, save for $s = 1/2$, $\lambda = 1/4$. For $s = 1/2$ the above eigenfunctions become

$$(1.24) \qquad\qquad y^{1/2} \qquad \text{and} \qquad y^{1/2}\log y,$$

respectively. If $s \neq 1/2$, we shall often take a simpler pair y^s, y^{1-s}.

If one wants $f(z)$ to be periodic in x of period 1, try $f(z) = e(x)\,F(2\pi y)$ and find that F satisfies the ordinary differential equation

$$(1.25) \qquad\qquad F''(y) + (\lambda y^{-2} - 1)\,F(y) = 0.$$

There are two linearly independent solutions, namely

$$(2\pi^{-1}y)^{1/2}K_{s-1/2}(y) \sim e^{-y}$$

and

$$(2\pi y)^{1/2}I_{s-1/2}(y) \sim e^y,$$

as $y \to +\infty$, where $K_\nu(y)$ and $I_\nu(y)$ are the standard Bessel functions (see Appendix B.4). Suppose that $f(z)$ does not grow too fast; more precisely,

$$f(z) = o(e^{2\pi y}),$$

as $y \to +\infty$. This condition rules out the second solution, so $f(z)$ is a multiple of the function

$$(1.26) \qquad\qquad W_s(z) = 2\,y^{1/2}K_{s-1/2}(2\pi y)\,e(x),$$

which is named the *Whittaker function*. It will be convenient to extend $W_s(z)$ to the lower half-plane $\overline{\mathbb{H}}$ by imposing the symmetry

$$(1.27) \qquad\qquad W_s(z) = W_s(\bar{z}).$$

We do not define $W_s(z)$ on the real line $\hat{\mathbb{R}} = \mathbb{R} \cup \{\infty\}$.

Now let us show how the Whittaker function evolves by the method of averaging images. For a function $f(z)$ to be periodic in x of period 1, one needs to verify the transformation rule

$$(1.28) \qquad\qquad f(nz) = \chi(n) f(z), \qquad \text{for all } n \in N,$$

where $\chi : N \longrightarrow \mathbb{C}$ is the character given by

$$\chi(n) = e(x), \qquad \text{if } n = \begin{pmatrix} 1 & x \\ & 1 \end{pmatrix}.$$

To construct such an f, we begin with the obvious eigenfunction y^s, by means of which we produce $\bar{\chi}(n)(\operatorname{Im} \omega n z)^s$, and integrate these over the group N, getting

$$
\begin{aligned}
f(z) &= \int_{-\infty}^{+\infty} \bar{\chi}(n(u)) \, (\operatorname{Im} \omega n(u) z)^s \, du \\
&= \int_{-\infty}^{+\infty} e(u) \left(\operatorname{Im} \frac{-1}{z-u} \right)^s du \\
&= e(x) \, y^{1-s} \int_{-\infty}^{+\infty} (1+t^2)^{-s} e(ty) \, dt = \pi^s \Gamma(s)^{-1} W_s(z).
\end{aligned}
$$

Here the involution $\omega = \begin{pmatrix} & -1 \\ 1 & \end{pmatrix}$ was inserted to buy the absolute convergence, at least if $\operatorname{Re} s > 1/2$. The resulting function has an analytic continuation to the whole complex s-plane, where it extends to an eigenfunction of Δ, periodic in x.

Changing the character into $\chi(n) = e(rx)$, where r is a fixed real number different from 0, we obtain by the above method an eigenfunction which is a multiple of $W_s(rz)$. The Whittaker functions are basic for harmonic analysis on $\mathbb{H}$, as the following proposition clearly assures.

Proposition 1.4. *Any* $f \in C_0^\infty(\mathbb{H})$ *has the integral representation*

$$(1.29) \qquad\qquad f(z) = \frac{1}{2\pi i} \int_{(1/2)} \int_{\mathbb{R}} W_s(rz) \, f_s(r) \, \gamma_s(r) \, ds \, dr,$$

where the outer integration is taken over the vertical line $s = 1/2 + it$,

$$(1.30) \qquad f_s(r) = \int_{\mathbb{H}} f(z) \, W_s(rz) \, d\mu z,$$

and $\gamma_s(r) = (2\pi|r|)^{-1} t \sinh \pi t$.

Therefore, loosely speaking, the Whittaker functions $W_s(rz)$ with $\operatorname{Re} s = 1/2$ and $r \neq 0$ real form a complete eigenpacket on $\mathbb{H}$. The proof of Proposition 1.4 is obtained by application of the classical Fourier inversion in (r, x) variables and the following Kontorovitch-Lebedev inversion in (t, y) variables (see [Ko-Le] and [Le, p.131]):

$$(1.31) \qquad g(w) = \int_0^{+\infty} K_{it}(w) \left(\int_0^{+\infty} K_{it}(y) \, g(y) \, y^{-1} \, dy \right) \pi^{-2} t \, \sinh(\pi t) \, dt.$$

The proof of the next result reduces to the ordinary Fourier series expansion for periodic functions.

Proposition 1.5. *Let $f(z)$ be an eigenfunction of Δ with eigenvalue $\lambda = s(1 - s)$ which satisfies the transformation rule*

$$(1.32) \qquad f(z + m) = f(z) \qquad \text{for all } m \in \mathbb{Z}$$

and the growth condition

$$(1.33) \qquad f(z) = o(e^{2\pi y}) \qquad \text{as } y \to +\infty.$$

Then $f(z)$ has the expansion

$$(1.34) \qquad f(z) = f_0(y) + \sum_{n \neq 0} f_n \, W_s(nz),$$

where the zero-term $f_0(y)$ is a linear combination of the functions in (1.23) and (1.24). The series converges absolutely and uniformly on compacta. Hence

$$(1.35) \qquad f_n \ll e^{\varepsilon|n|}$$

for any $\varepsilon > 0$, with the implied constant depending on f and ε.

No less important than $W_s(z)$ is the eigenfunction of Δ associated with the second solution to the differential equation (1.25), given by

$$(1.36) \qquad V_s(z) = 2\pi \, y^{1/2} I_{s-1/2}(2\pi y) \, e(x).$$

We extend $V_s(z)$ to the lower half-plane $\overline{\mathbb{H}}$ by requiring the same symmetry as for $W_s(z)$, and we do not define $V_s(z)$ on the real line. Note that $W_s(z)$ and $V_s(z)$ have distinct behaviour at infinity, namely

$$(1.37) \qquad\qquad W_s(z) \sim e(x)\, e^{-2\pi y},$$
$$(1.38) \qquad\qquad V_s(z) \sim e(x)\, e^{2\pi y},$$

as $y \to +\infty$; therefore, they are linearly independent. They both will appear in the Fourier expansion of the automorphic Green function.

Next we shall perform the harmonic analysis on $\mathbb{H}$ in geodesic polar coordinates (r, φ). Recall the connection $z = k(\varphi)\, e^{-r} i$. We seek an eigenfunction of Δ with eigenvalue $\lambda = s(1-s)$ which transforms as

$$(1.39) \qquad\qquad f(kz) = \chi(k)\, f(z)\,, \qquad \text{for all } k \in K\,,$$

where $\chi : K \longrightarrow \mathbb{C}$ is the character given by (for $m \in \mathbb{Z}$)

$$\chi(k) = e^{2im\theta}\,, \qquad \text{if } k = \begin{pmatrix} \cos\theta & \sin\theta \\ -\sin\theta & \cos\theta \end{pmatrix}\,.$$

To produce such an eigenfunction we integrate $\bar{\chi}(k)(\operatorname{Im} kz)^s$ over the group K, getting

$$\begin{aligned} f(z) &= \frac{1}{\pi} \int_0^\pi (\operatorname{Im} k(-\theta)\, k(\varphi)\, e^{-r} i)^s e^{2im\theta}\, d\theta \\ &= \frac{1}{\pi} \int_0^\pi (\cosh r + \sinh r \,\cos 2\theta)^{-s} e^{2im(\theta+\varphi)}\, d\theta \\ &= \frac{\Gamma(1-s)}{\Gamma(1-s+m)}\, P_{-s}^m(\cosh r)\, e^{2im\varphi}\,, \end{aligned}$$

where $P_{-s}^m(v)$ is the Legendre function (the gamma factors have appeared because of unfortunate normalization in the literature on special functions, see Appendix B.3).

The same eigenfunction can be obtained by the method of separation of variables. Indeed, writing $f(z) = F(u)\, e^{2im\varphi}$, where $2u + 1 = \cosh r$, we find by (1.21) that $F(u)$ solves the ordinary differential equation

$$u(u+1)\, F''(u) + (2u+1)\, F'(u) + \left(s(1-s) - \frac{m^2}{4u(u+1)} \right) F(u) = 0\,.$$

Then we verify by partial integration that $F(u)$ given by

$$F(u) = \frac{1}{\pi} \int_0^\pi (2u + 1 + 2\sqrt{u(u+1)}\,\cos\theta)^{-s} \cos(m\theta)\, d\theta$$

$$= \frac{\Gamma(1-s)}{\Gamma(1-s+m)} \, P^m_{-s}(2u+1)$$

is a solution to this equation (see (B.21) and (B.24)).

By either method we have obtained the classical spherical functions

$$(1.40) \qquad\qquad U^m_s(z) = P^m_{-s}(2u+1) \, e^{2im\varphi} \, .$$

They form a complete system on $\mathbb{H}$ in the following sense.

Proposition 1.6. *Any* $f \in C^\infty_0(\mathbb{H})$ *has the expansion*

$$(1.41) \qquad\qquad f(z) = \sum_{m\in\mathbb{Z}} \frac{(-1)^m}{2\pi i} \int_{(1/2)} U^m_s(z) \, f^m(s) \, \delta(s) \, ds \, ,$$

where the integration is taken over the vertical line $s = 1/2 + it$,

$$f^m(s) = \int_{\mathbb{H}} f(z) \, U^m_s(z) \, d\mu z \, ,$$

and $\delta(s) = t \tanh \pi t$.

The above expansion can be derived by applications of the Fourier series representation of a periodic function together with the following inversion formula due to F. G. Mehler [Me] and V. A. Fock [Foc]:

$$(1.42) \qquad g(u) = \int_0^{+\infty} P_{-1/2+it}(u) \left(\int_1^{+\infty} P_{-1/2+it}(v) \, g(v) \, dv \right) t \tanh(\pi t) \, dt \, .$$

Here $P_s(v) = P^0_s(v)$ denotes the Legendre function of order $m = 0$.

The spherical functions of order zero are special; they depend only on the hyperbolic distance, namely $U^0_s(z) = P_{-s}(2u+1) = F_s(u)$, say. Note that

$$(1.43) \qquad\qquad F_s(u) = \frac{1}{\pi} \int_0^{\pi} (2u + 1 + 2\sqrt{u(u+1)} \cos\theta)^{-s} \, d\theta$$

is also given by the hypergeometric function (see (B.23))

$$F_s(u) = F(s, 1-s; 1, u) \, ,$$

and it satisfies the differential equation

$$(1.44) \qquad\qquad u(u+1) \, F''(u) + (2u+1) \, F'(u) + s(1-s) \, F(u) = 0 \, .$$

Recall that the differential equation (1.44) is equivalent to (see (1.21))

$$(1.45) \qquad (\Delta + s(1-s))\, F = 0$$

for functions depending only on the distance variable u.

There is another solution to (1.44), linearly independent of $F_s(u)$, given by

$$(1.46) \qquad G_s(u) = \frac{1}{4\pi} \int_0^1 (\xi(1-\xi))^{s-1}(\xi + u)^{-s}\, d\xi\,.$$

This also can be expressed by the hypergeometric function (see (B.16)):

$$G_s(u) = \frac{\Gamma(s)^2}{4\pi\,\Gamma(2s)}\, u^{-s}\, F\!\left(s, s; 2s, \frac{1}{u}\right).$$

Lemma 1.7. *The integral* (1.46) *converges absolutely for* $\operatorname{Re} s = \sigma > 0$. *It gives a function* $G_s(u)$ *on* $\mathbb{R}^+$ *which satisfies equations* (1.44) *and* (1.45). *Moreover,* $G_s(u)$ *satisfies the following bounds:*

$$(1.47) \qquad G_s(u) = \frac{1}{4\pi} \log \frac{1}{u} + O(1)\,, \qquad u \to 0\,,$$

$$(1.48) \qquad G_s'(u) = -(4\pi u)^{-1} + O(1)\,, \qquad u \to 0\,,$$

$$(1.49) \qquad G_s(u) \ll u^{-\sigma}\,, \qquad u \to +\infty\,.$$

Proof. That $G_s(u)$ satisfies (1.45) follows by partial integration from the identity

$$(\Delta + s(1-s))\,\xi^{s-1}(1-\xi)^{s-1}\xi^{-s} = s\frac{d}{d\xi}\,\xi^s(1-\xi)^s(\xi + u)^{s-1}\,.$$

To prove (1.47) we put $\nu = (|s| + 1)u$, $\eta = (|s| + 1)^{-1}$ and split

$$4\pi G_s(u) = \int_0^1 \left(\frac{\xi(1-\xi)}{\xi + u}\right)^{s-1} \frac{d\xi}{\xi + u} = \int_0^\nu + \int_\nu^\eta + \int_\eta^1$$

where

$$\int_0^\nu \ll u^{-\sigma} \int_0^\nu \xi^{\sigma-1}\, d\xi \ll 1$$

and

$$\int_\eta^1 \ll \int_\eta^1 (1-\xi)^{\sigma-1}\, d\xi \ll 1\,.$$

For the remaining integral we shall use the approximation

$$\left(\frac{\xi(1-\xi)}{\xi+u}\right)^{s-1} = \left(1 - \frac{u+\xi^2}{u+\xi}\right)^{s-1} = 1 + O\left(\frac{u+\xi^2}{u+\xi}\right)$$

and obtain

$$\int_{\nu}^{\eta} = \int_{\nu}^{\eta} \frac{d\xi}{\xi+u} + O\left(\int_{\nu}^{\eta} \frac{u+\xi^2}{(u+\xi)^2}\, d\xi\right)$$

$$= \log\frac{u+\eta}{u+\nu} + O(1) = \log\frac{1}{u} + O(1)\,.$$

This completes the proof of (1.47). The proof of (1.48) is similar, and (1.49) is obvious.

1.8. The invariant integral operators

An integral operator is defined by

$$(Lf)(z) = \int_{\mathbb{H}} k(z,w)\, f(w)\, d\mu w\,,$$

where $d\mu$ is the riemannian measure and $k : \mathbb{H} \times \mathbb{H} \longrightarrow \mathbb{C}$ is a given function, called the *kernel* of L. In what follows we always assume without mention that the kernel $k(z,w)$ and the test function $f(w)$ are such that the integral converges absolutely. This assumption does not exclude the possibility that $k(z,w)$ is singular; as a matter of fact the important kernels are singular on the diagonal $z = w$.

For L to be invariant it is necessary and sufficient that

$$k(gz, gw) = k(z,w)\,, \qquad \text{for all } g \in G\,.$$

A function with this property is called *point-pair invariant*; it depends solely on the hyperbolic distance between the points. Consequently, we can set

$$k(z,w) = k(u(z,w))\,,$$

where $k(u)$ is a function in one variable $u \geq 0$ and $u(z,w)$ is given by (1.4). Therefore, an invariant integral operator is a convolution.

The invariant integral operators will be used to develop the spectral resolution of the Laplace operator. The key point is that the resolvent of Δ (the inverse to $\Delta + s(1-s)$ acting on functions satisfying suitable growth conditions) is an integral operator with kernel $G_s(u)$ given by (1.46). On the other hand, every invariant integral operator is a function of Δ in a spectral sense. We shall give a proof of this important fact first because it requires several independent results, which will be employed elsewhere.

Lemma 1.8. *Let $k(z, w)$ be a smooth point-pair invariant on $\mathbb{H} \times \mathbb{H}$. We have*

$$(1.50) \qquad \Delta_z k(z, w) = \Delta_w k(z, w) \,.$$

Proof. Using geodesic polar coordinates with the origin at w (send i to w), we get

$$\Delta_z k(z, w) = u(u + 1) \, k''(u) + (2u + 1) \, k'(u) \,.$$

Then, using geodesic polar coordinates with the origin at z, we get the same expression for $\Delta_w k(z, w)$.

For two functions F, G such that $|FG|$ is integrable on $\mathbb{H}$ with respect to the measure $d\mu$, we define the inner product by

$$(1.51) \qquad \langle F, G \rangle = \int_{\mathbb{H}} F(z) \, \overline{G}(z) \, d\mu z \,.$$

If $F, G \in C_0^\infty(\mathbb{H})$, then by partial integration

$$(1.52) \qquad \langle -\Delta F, G \rangle = \int_{\mathbb{H}} \nabla F \cdot \overline{\nabla G} \, dx \, dy \,,$$

where $\nabla F = [\partial F/\partial x, \partial F/\partial y]$ is the gradient of F. Hence we infer that

$$(1.53) \qquad \langle -\Delta F, G \rangle = \langle F, -\Delta G \rangle$$

and

$$(1.54) \qquad \langle -\Delta F, F \rangle \geq 0 \,.$$

Hence, $-\Delta$ is a symmetric and non-negative operator in the space $C_0^\infty(\mathbb{H})$.

Theorem 1.9. *The invariant integral operators commute with the Laplace operator.*

Proof. By (1.50) and (1.53) we argue as follows:

$$\Delta L \, f(z) = \int \Delta_z k(z, w) \, f(w) \, d\mu w = \int \Delta_w k(z, w) \, f(w) \, d\mu w$$
$$= \int k(z, w) \Delta_w f(w) \, d\mu w = L \, \Delta \, f(z).$$

Remarks. The lower bound (1.54) can be improved and generalized a bit. Consider the Dirichlet problem

$$(1.55) \qquad \begin{cases} (\Delta + \lambda) \, F = 0 & \text{in } D\,, \\ \quad\;\; F = 0 & \text{on } \partial D \end{cases}$$

for a domain $D \subset \mathbb{H}$ with a piecewise continuous boundary ∂D, where F is smooth in D and continuous in $\partial D \cup D$. The solutions are in the Hilbert space with respect to the inner product (1.51), where $\mathbb{H}$ is reduced to D. Observe that the formula (1.52) remains valid if $\mathbb{H}$ is replaced by D. This yields the following inequality:

$$\langle -\Delta F, F \rangle = \int_D \left(\left(\frac{\partial F}{\partial x} \right)^2 + \left(\frac{\partial F}{\partial y} \right)^2 \right) dx\, dy \geq \int_D \left(\frac{\partial F}{\partial y} \right)^2 dx\, dy.$$

On the other hand, by partial integration we have, for each x,

$$\int F^2 \, y^{-2} \, dy = 2 \int F \frac{\partial F}{\partial y} \, y^{-1} \, dy,$$

and integrating in x we infer, by the Cauchy-Schwarz inequality, that

$$\int_D F^2 \, d\mu \leq 4 \int_D \left(\frac{\partial F}{\partial y} \right)^2 dx\, dy.$$

Combining these two estimates, we obtain

$$(1.56) \qquad\qquad \langle -\Delta F, F \rangle \geq \frac{1}{4} \langle F, F \rangle$$

(we have assumed tacitly that F is real, but, of course, this is not necessary). Hence we conclude that if F is a non-zero solution to the Dirichlet problem for a domain in the hyperbolic plane, then its eigenvalue satisfies $\lambda \geq 1/4$. This fact explains the absence of Whittaker functions $W_s(rz)$ in (1.29) and the spherical functions $U_s^m(z)$ in (1.41) off the line $\mathrm{Re}\, s = 1/2$.

We return to the study of point-pair invariants. A function $f(z, w)$ is said to be *radial at w* if as a function of z it depends only on the distance of z to w, *i.e.* it can be written as $F(u(z, w), w)$. A function $f(z, w)$ can be radial at some point w, but not necessarily at other points; clearly a point-pair invariant is radial at all points. Given any $f : \mathbb{H} \longrightarrow \mathbb{C}$, one can produce a radial function at $w \in \mathbb{H}$ by averaging over the stability group $G_w = \{ g \in G : \; gw = w \}$. One gets

$$f_w(z) = \int_{G_w} f(gz) \, dg = \frac{1}{\pi} \int_0^\pi f(\sigma \, k(\theta) \, \sigma^{-1}) \, d\theta\,,$$

where $\sigma \in G$ is any motion which brings i to w so that $G_w = \sigma K \sigma^{-1}$. The mapping $f \mapsto f_w$ will be called *the mean-value operator*.

Lemma 1.10. *The mean-value $f_w(z)$ is radial at w. Moreover,*

$$(1.57) \qquad\qquad f_z(z) = f(z) \,.$$

Proof. Suppose z, z_1 are at the same distance from w. Then there exists $g_1 \in G_w$ which sends z_1 to z. Applying g_1, we derive that

$$f_w(z_1) = \int_{G_w} f(gz_1)\,dg = \int_{G_w} f(gg_1z)\,dg = \int_{G_w} f(gz)\,dg = f_w(z),$$

thus proving the first assertion. The second assertion is straightforward:

$$f_z(z) = \int_{G_z} f(gz)\,dg = f(z) \int_{G_w} dg = f(z)\,.$$

Lemma 1.11. *An invariant integral operator L is not altered by the mean-value operator, i.e. we have*

$$(Lf)(z) = (Lf_z)(z)\,.$$

Proof. Let $k(z,w)$ be a kernel of L which is point-pair invariant. We argue as follows:

$$(Lf_z)(z) = \int_{\mathbb{H}} k(z,w)\,f_z(w)\,d\mu w = \int_{\mathbb{H}} k(z,w)\left(\int_{G_z} f(gw)\,dg \right) d\mu w$$

$$= \int_{G_z} \left(\int_{\mathbb{H}} k(z,w)\,f(gw)\,d\mu w \right) dg = \int_{G_z} \left(\int_{\mathbb{H}} k(gz,w)\,f(w)\,d\mu w \right) dg$$

$$= \left(\int_{G_z} dg \right)\left(\int_{\mathbb{H}} k(z,w)\,f(w)\,d\mu w \right) = (Lf)(z)\,.$$

A function $f(z,w)$ which is radial at every $w \in \mathbb{H}$ may not necessarily be a point-pair invariant, but if in addition $f(z,w)$ is an eigenfunction of Δ in z for any w with eigenvalue independent of w, then $f(z,w)$ is a point-pair invariant. Such a function is unique up to a constant factor.

Lemma 1.12. *Let $\lambda \in \mathbb{C}$ and $w \in \mathbb{H}$. There exists a unique function $\omega(z,w)$ in z which is radial at w such that*

$$\omega(w,w) = 1\,,$$
$$(\Delta_z + \lambda)\,\omega(z,w) = 0\,.$$

This is given by

$$(1.58) \qquad\qquad \omega(z,w) = F_s(u(z,w))\,,$$

where $F_s(u)$ is the Gauss hypergeometric function $F(s, 1-s; 1; u)$.

Proof. Setting $\omega(z, w) = F(u)$ with $u = u(z, w)$, we find that $F(u)$ satisfies the differential equation (1.44); thus it is a linear combination of $F_s(u)$ and $G_s(u)$, but the normalization condition $F(0) = \omega(w, w) = 1$ determines (1.58).

It follows immediately from Lemmas 1.10 and 1.12 that

Corollary 1.13. *If $f(z)$ is an eigenfunction of Δ with eigenvalue $\lambda = s(1 - s)$, then*

$$(1.59) \qquad f_w(z) = \omega(z, w) \, f(w) \,.$$

As a consequence, notice that if an eigenfunction f vanishes at a point w, then $f_w \equiv 0$. Now we are ready to prove the following basic result.

Theorem 1.14. *Any eigenfunction of Δ is also an eigenfunction of all invariant integral operators. More precisely, if $(\Delta + \lambda)f = 0$ and L is an integral operator whose kernel $k(u)$ is smooth and compactly supported in $\mathbb{R}^+$, then there exists $\Lambda = \Lambda(\lambda, k) \in \mathbb{C}$, depending on λ and k but not on f, such that $Lf = \Lambda \cdot f$, i.e.*

$$(1.60) \qquad \int_{\mathbb{H}} k(z, w) \, f(w) \, d\mu w = \Lambda \, f(z) \,.$$

Proof. By Lemma 1.11 and Corollary 1.13 we obtain (1.60) with

$$(1.61) \qquad \Lambda = \int_{\mathbb{H}} k(z, w) \, \omega(z, w) \, d\mu w \,.$$

It remains to show that the above integral does not depend on z. But this is obvious, because G acts on $\mathbb{H}$ transitively and ω, k are point-pair invariants.

The converse to Theorem 1.14 is also true. It asserts the following:

Theorem 1.15. *If f is an eigenfunction of all invariant integral operators whose kernel functions are in $C_0^\infty(\mathbb{R}^+)$, then f is an eigenfunction of Δ.*

Proof. Let $k(z, w)$ be a point-pair invariant such that (1.60) holds true with $\Lambda \neq 0$ (if $\Lambda = 0$ for all k, then $f \equiv 0$ and the assertion is obvious). Applying Δ to both sides, we get

$$\int_{\mathbb{H}} \Delta_z k(z, w) \, f(w) \, d\mu w = \Lambda (\Delta f)(z) \,.$$

But $\Delta_z k(z, w)$ is another point-pair invariant, so by the hypothesis the above integral equals $\Lambda' f(z)$ for some $\Lambda' \in \mathbb{C}$. By combining the two relations, we get $(\Delta + \lambda)f = 0$ with $\lambda = -\Lambda'\Lambda^{-1}$.

There is a striking resemblance between Cauchy's formula for holomorphic functions and the integral representation (1.60) for eigenfunctions of Δ. The latter is particularly helpful for testing the convergence of sequences of eigenfunctions, as well as for estimating at individual points.

For applications we need an explicit expression for Λ in terms of the eigenvalue λ and the kernel function $k(u)$. This is given by the Selberg/Harish-Chandra transform in the following three steps:

(1.62)
$$q(v) = \int_v^{+\infty} k(u) (u - v)^{-1/2} \, du \,,$$

$$g(r) = 2 q \left(\left(\sinh \frac{r}{2} \right)^2 \right),$$

$$h(t) = \int_{-\infty}^{+\infty} e^{irt} g(r) \, dr.$$

Theorem 1.16. *If $k \in C_0^\infty(\mathbb{R}^+)$ and if f is an eigenfunction of Δ with eigenvalue $\lambda = s(1 - s)$, where $s = 1/2 + it$, $t \in \mathbb{C}$, then (1.60) holds with $\Lambda = h(t)$.*

Proof. Since Λ does not depend on the eigenfunction $f(z)$, we take for computation $f(w) = (\operatorname{Im} w)^s$ and specialize (1.60) to the point $z = i$, getting

$$\Lambda = 2 \int_0^{+\infty} \int_0^{+\infty} k \left(\frac{x^2 + (y - 1)^2}{4 y} \right) y^{s-2} \, dx \, dy \,.$$

Changing the variable $x = 2\sqrt{uy}$ and next the variable $y = e^r$, one easily completes the computation, getting $\Lambda = h(t)$.

Theorem 1.16 says that an invariant integral operator is a function of the Laplace operator (in the spectral sense) given by the Selberg/Harish-Chandra transform (1.62). The assumption in Theorems 1.14-1.16 that the kernel $k(u)$ is compactly supported is not essential, though a certain control over the growth is required. It is simpler to express the sufficient conditions in terms of $h(t)$ rather than $k(u)$. These conditions are:

(1.63)
$$h(t) \text{ is even},$$

$$h(t) \text{ is holomorphic in the strip } |\operatorname{Im} t| \leq \frac{1}{2} + \varepsilon \,,$$

$$h(t) \ll (|t| + 1)^{-2-\varepsilon} \text{ in the strip.}$$

For any h having the above properties, one finds the inverse of the Selberg/Harish-Chandra transform in the following three steps:

$$g(r) = \frac{1}{2\pi} \int_{-\infty}^{+\infty} e^{irt} h(t) \, dt \,,$$

(1.64)
$$q(v) = \frac{1}{2} g(2 \log(\sqrt{v+1} + \sqrt{v})) \,,$$

$$k(u) = -\frac{1}{\pi} \int_u^{+\infty} (v - u)^{-1/2} \, dq(v) \,.$$

We shall rarely apply the relations (1.62) and (1.64), since they are quite knotty. Instead, it is often easier to assess $h(t)$ from the integral representation (1.60) by testing it against a suitable eigenfunction. For $f(w) = \omega(z, w) = F_s(u(z, w))$ we get by (1.61) (using polar coordinates (u, φ) and (1.17)) that

(1.62')
$$h(t) = 4\pi \int_0^{+\infty} F_s(u) \, k(u) \, du \,,$$

where $s = 1/2 + it$ and $F_s(u)$ is the Gauss hypergeometric function (see (1.43) and (1.44)). The inverse is given by (applying (1.42))

(1.64')
$$k(u) = \frac{1}{4\pi} \int_{-\infty}^{+\infty} F_s(u) \, h(t) \, \tanh(\pi t) \, t \, dt \,.$$

1.9. The Green function on $\mathbb{H}$

Let $s \in \mathbb{C}$ with $\operatorname{Re} s > 1$, and let $-R_s$ be the integral operator on $\mathbb{H}$ whose kernel function is given by (1.46), *i.e.*

(1.65)
$$-(R_s f)(z) = \int_{\mathbb{H}} G_s(u(z, w)) \, f(w) \, d\mu w \,.$$

Theorem 1.17. *If f is smooth and bounded on $\mathbb{H}$, then*

(1.66)
$$(\Delta + s(1 - s)) R_s f = f \,.$$

In other words, R_s is the right inverse to $\Delta + s(1-s)$, so that $G_s(u(z, w))$ is the Green function on the free space $\mathbb{H}$. Recall that

(1.67)
$$(\Delta + s(1 - s)) G_s = 0 \,.$$

Before proving Theorem 1.17, let us make a few remarks. First we emphasize that G_s is singular on the diagonal $z = w$. More precisely, we have

(1.68)
$$G_s(u(z, w)) = \frac{-1}{2\pi} \log |z - w| + H_s(z, w),$$

say, where H_s is smooth and has bounded derivatives on $\mathbb{H} \times \mathbb{H}$. The logarithmic singularity of G_s is the critical property for (1.66) to hold true for any smooth f; indeed, ignoring this property, one could guess wrongly that the operator $\Delta + s(1-s)$ annihilates R_s, since it annihilates G_s.

The proof of (1.66) depends on the following formula.

Lemma 1.18. *If f is smooth and bounded on $\mathbb{H}$, then*

$$(1.69) \quad -(\Delta+s(1-s))R_sf(z) = \int_{\mathbb{H}} G_s(u(z,w))\,(\Delta+s(1-s))f(w)\,d\mu w\,.$$

A formal argument using the symmetry of $\Delta + s(1-s)$ (which is not justified for singular kernels, because an integration by parts makes the derivative of the kernel not integrable) seems to yield the result immediately, but a rigorous proof is quite subtle . To this end we use some differential operators derived from the Lie algebra of $G = SL_2(\mathbb{R})$. Let us recall a few basic facts (*cf.* [La]). The Lie algebra $\mathfrak{g}$ of the group G over $\mathbb{R}$ consists of all 2×2 matrices X such that

$$\exp{(tX)} = \sum_{n=0}^{\infty} \frac{(tX)^n}{n!} \in G\,, \qquad \text{for all } t \in \mathbb{R}\,.$$

One can show that $\mathfrak{g}$ consists of trace zero matrices and that

$$X_1 = \begin{pmatrix} 0 & 1 \\ 0 & 0 \end{pmatrix}, \qquad X_2 = \begin{pmatrix} 0 & 0 \\ 1 & 0 \end{pmatrix}, \qquad X_3 = \begin{pmatrix} 1 & 0 \\ 0 & -1 \end{pmatrix}$$

form a basis of $\mathfrak{g}$ over $\mathbb{R}$. Note that for the basis matrices we have

$$\exp{(tX_1)} = 1 + tX_1 = \begin{pmatrix} 1 & t \\ & 1 \end{pmatrix}\,,$$

$$\exp{(tX_2)} = 1 + tX_2 = \begin{pmatrix} 1 & \\ t & 1 \end{pmatrix}\,,$$

$$\exp{(tX_3)} = 1 + tX_3 + \frac{t^2}{2} + \frac{t^3}{6}X_3 + \cdots = \begin{pmatrix} e^t & \\ & e^{-t} \end{pmatrix}\,.$$

If $X \in \mathfrak{g}$, then $\{\exp{(tX)} : t \in \mathbb{R}\}$ is a one-parameter subgroup of G, and the map $t \mapsto \exp{(tX)}$ is a group homomorphism which is real-analytic in a neigbourhood of $t = 0$. Thus we can define a linear operator $\mathcal{L}_X : C^\infty(G) \longrightarrow C^\infty(G)$ by

$$(1.70) \qquad\qquad \mathcal{L}_Xf(z) = \frac{d}{dt}f(\exp{(tX)}z)\Big|_{t=0}\,.$$

Clearly $\mathcal{L}_X$ satisfies the Leibniz rule $\mathcal{L}_X(fg) = f\mathcal{L}_X g + g\mathcal{L}_X f$, and so $\mathcal{L}_X$ is a differentiation (the Lie derivative).

Let $\mathcal{L}_1$, $\mathcal{L}_2$, $\mathcal{L}_3$ denote the differential operators derived from the basis matrices X_1, X_2, X_3 respectively. We shall show that

$$(1.71) \qquad \mathcal{L}_1 = \frac{\partial}{\partial x}\,,$$

$$(1.72) \qquad \mathcal{L}_2 = (y^2 - x^2)\frac{\partial}{\partial x} - 2xy\frac{\partial}{\partial y}\,,$$

$$(1.73) \qquad \mathcal{L}_3 = 2x\frac{\partial}{\partial x} + 2y\frac{\partial}{\partial y}\,,$$

$$(1.74) \qquad 2\Delta = \mathcal{L}_1\mathcal{L}_2 + \mathcal{L}_2\mathcal{L}_1 + \frac{1}{2}\mathcal{L}_3\mathcal{L}_3\,.$$

The formula (1.71) is obtained by differentiating at $t = 0$ as follows:

$$\mathcal{L}_1 f(z) = \frac{d}{dt}f(z + t) = f_x(z)\,.$$

The formula (1.72) is obtained by differentiating at $t = 0$ as follows:

$$\mathcal{L}_2 f(z) = \frac{d}{dt}f\Big(\frac{z}{tz + 1}\Big) = \frac{d}{dt}\Big(\mathrm{Re}\,\frac{z}{tz + 1}\Big)f_x(z) + \frac{d}{dt}\Big(\mathrm{Im}\,\frac{z}{tz + 1}\Big)f_y(z)\,,$$

$$\frac{d}{dt}\Big(\frac{z}{tz + 1}\Big) = -z^2 = -x^2 + y^2 - 2i\,xy\,.$$

The formula (1.73) is obtained by differentiating at $t = 0$ as follows:

$$\mathcal{L}_3 f(z) = \frac{d}{dt}f(e^{2t}z) = 2xf_x(z) + 2yf_y(z)\,.$$

Finally, (1.74) is obtained by adding the following easy formulas:

$$\mathcal{L}_1\mathcal{L}_2 = (y^2 - x^2)\frac{\partial^2}{\partial x^2} - 2xy\frac{\partial^2}{\partial x\partial y} - 2x\frac{\partial}{\partial x} - 2y\frac{\partial}{\partial y}\,,$$

$$\mathcal{L}_2\mathcal{L}_1 = (y^2 - x^2)\frac{\partial^2}{\partial x^2} - 2xy\frac{\partial^2}{\partial x\partial y}\,,$$

$$\frac{1}{2}\mathcal{L}_3\mathcal{L}_3 = 2x^2\frac{\partial^2}{\partial x^2} + 4xy\frac{\partial^2}{\partial x\partial y} + 2y^2\frac{\partial^2}{\partial y^2} + 2x\frac{\partial}{\partial x} + 2y\frac{\partial}{\partial y}\,.$$

Now we are ready to give a rigorous proof of (1.69). Let $g_t = \exp(tX_j)$. Since R_s is a G-invariant operator, we have

$$-R_s f(g_t z) = \int_{\mathbb{H}} G_s(u(z, w))\,f(g_t w)\,d\mu w\,.$$

Differentiating with respect to t and then putting $t = 0$, we get (1.69) for each of the three operators $\mathcal{L}_j$ in place of $\Delta + s(1-s)$; hence (1.69) is derived for $\Delta + s(1-s)$ by (1.74).

For the proof of Theorem 1.17 we split $\mathbb{H} = U \cup V$, where U is the disc (euclidean) centered at z of radius $\varepsilon > 0$ and V is the area outside the disc; accordingly we split the integral (1.69). One sees clearly that the integral over the disc U vanishes as ε tends to 0. To evaluate the complementary integral over V we shall use Green's formula

$$(1.75) \qquad \int_V (g\,Df - f\,Dg)\,dx\,dy = \int_{\partial V} \left(g\,\frac{\partial f}{\partial n} - f\,\frac{\partial g}{\partial n} \right) d\ell \,,$$

where $D = \partial^2/\partial x^2 + \partial^2/\partial y^2$ is the Laplace operator on $\mathbb{R}^2$, $\partial/\partial n$ is the outer normal derivative and $d\ell$ is the euclidean length element. We get

$$\int_V G(u(z,w))\,(\Delta + s(1-s))f(w)\,d\mu w = \int_{\partial U} \left(G\,\frac{\partial f}{\partial n} - f\,\frac{\partial G}{\partial n} \right) d\ell$$

by (1.67). Here on the right-hand side the integral of $G\,\partial f/\partial n$ vanishes as $\varepsilon \to 0$, so we are left with

$$\int_{\partial U} f\,\frac{\partial G}{\partial n}\,d\ell = -\frac{1}{2\pi}\int_{\partial U} f(w)\,\frac{\partial \log|z-w|}{\partial n}\,d\ell + \int_{\partial U} f(w)\,\frac{\partial H(z,w)}{\partial n}\,d\ell$$

by (1.68). The last integral vanishes as $\varepsilon \to 0$, and the preceding one is equal to (using euclidean polar coordinates)

$$\frac{1}{2\pi\varepsilon}\int_{\partial U} f(w)\,d\ell \,.$$

This tends to $f(z)$ as $\varepsilon \to 0$, thus completing the proof of (1.66).

Chapter 2

Fuchsian Groups

In this chapter we give basic facts about groups of motions acting discontinuously on the hyperbolic plane.

2.1. Definitions

The group $M_2(\mathbb{R})$ of 2×2 real matrices is a vector space with inner product defined by

$$\langle g, h \rangle = \text{Trace}\, (gh^t) = \text{Tr}\, (gh^t).$$

One easily checks that $\|g\| = \langle g, g \rangle^{1/2}$ is a norm in $M_2(\mathbb{R})$ and that

$$\|g\|^2 = a^2 + b^2 + c^2 + d^2, \qquad \text{for} \quad g = \begin{pmatrix} a & b \\ c & d \end{pmatrix}.$$

Besides its usual properties, this norm satisfies $\|gh\| \leq \|g\|\,\|h\|$.

The embedding $G = SL_2(\mathbb{R}) \longrightarrow M_2(\mathbb{R})$ induces a metric topology in G. A subgroup $\Gamma \subset G$ is *discrete* if the induced topology in Γ is discrete, *i.e.* the set

$$\big\{ \gamma \in \Gamma : \ \|\gamma\| < r \big\}$$

is finite for any $r > 0$. Observe that a discrete group is countable.

Let X be a topological space (Hausdorff), and let Γ be a group of homeomorphisms of X acting on X. We say that Γ *acts on X discontinuously* if the orbit Γx of any $x \in X$ has no limit point in X. Equivalently, any compact subset $Y \subset X$ is disjoint with γY for all but a finite number of $\gamma \in \Gamma$. Observe that the stability group Γ_x of a point $x \in X$ is finite.

Proposition 2.1 (Poincaré). *A subgroup of $SL_2(\mathbb{R})$ is discrete if and only if when considered as a subgroup of $PSL_2(\mathbb{R})$ it acts discontinuously on $\mathbb{H}$.*

A subgroup $\Gamma \subset PSL_2(\mathbb{R})$ acting on $\mathbb{H}$ discontinuously is called a *Fuchsian group*. Naturally a Fuchsian group acts on $\hat{\mathbb{C}}$, but of course, it may not act discontinuosly on the whole $\hat{\mathbb{C}}$.

There is a multitude of Fuchsian groups. Surprising is the following result of J. Nielsen [Ni]: if $\Gamma \subset PSL_2(\mathbb{R})$ is non-abelian and hyperbolic (it contains only hyperbolic elements and the identity), then Γ acts discontinuously on $\mathbb{H}$ (an elegant proof was given by C. L. Siegel [Si1]).

Proposition 2.2. *Let Γ be a Fuchsian group and $z \in \hat{\mathbb{C}}$. Then the stability group*

$$\Gamma_z = \{\gamma \in \Gamma : \ \gamma z = z\}$$

is cyclic (not necessarily finite if $z \in \hat{\mathbb{R}}$).

An element γ_0 of a Fuchsian group is called *primitive* if γ_0 generates the stability group of the set of its fixed points, and in case γ_0 is elliptic it has the smallest angle of rotation. Any γ other than the identity motion is a power of a unique primitive element, $\gamma = \gamma_0^n$, $n \in \mathbb{Z}$.

A Fuchsian group Γ is said to be of the *first kind* if every point on the boundary $\partial\mathbb{H} = \hat{\mathbb{R}}$ is a limit (in the $\hat{\mathbb{C}}$-topology) of the orbit Γz for some $z \in \mathbb{H}$. Clearly, any subgroup of finite index of a Fuchsian group of the first kind is a Fuchsian group of the first kind. But a Fuchsian group of the first kind cannot be too small. Obviously, it cannot be cyclic, and a fortiori the stability group Γ_z of a point $z \in \hat{\mathbb{C}}$ is not of the first kind.

2.2. Fundamental domains

A Fuchsian group can be visualized by its fundamental domain. Two points $z, w \in \hat{\mathbb{C}}$ are said to be *equivalent* if $w \in \Gamma z$; we then write $z \equiv w \pmod{\Gamma}$. A set $F \subset \mathbb{H}$ is called a *fundamental domain for* Γ if

    i) F is a domain in $\mathbb{H}$,

    ii)   distinct points in F are not equivalent,

    iii)   any orbit of Γ contains a point in $\overline{F}$ (the closure of F in the $\hat{\mathbb{C}}$-topology).

Any Fuchsian group has a fundamental domain, not, of course, unique. However, all fundamental domains have the same positive volume (possibly infinite)

$$|F| = \int_F d\mu z.$$

A fundamental domain of a Fuchsian group Γ of the first kind can be chosen as a convex polygon. Specifically, suppose $w \in \mathbb{H}$ is not fixed by any $\gamma \in \Gamma$

other than the identity motion; then the set

$$D(w) = \big\{ z \in \mathbb{H} : \ \rho(z, w) < \rho(z, \gamma w) \ \text{ for all } \gamma \in \Gamma, \gamma \neq 1 \big\}$$

is a fundamental domain of Γ; it is called a *normal polygon* (due to Dirichlet). One can show that $D(w)$ is a polygon (connected and convex) with an even number of sides (subject to the convention that if a side contains a fixed point of an elliptic motion of order 2 from Γ, then this point is considered as a vertex and the side divided into two sides). The sides of $D(w)$ can be arranged in pairs of equivalent sides so that the side-pairing motions generate the group Γ.

From the above properties of $D(w)$ follows (for a complete proof see C. L. Siegel [Si2])

Proposition 2.3. *Every Fuchsian group of the first kind has a finite number of generators and a fundamental domain of finite volume.*

Conversely, one can show that if a Fuchsian group has fundamental domain of finite volume, then this group is of the first kind. For this reason a Fuchsian group of the first kind will be called more briefly a *finite volume group*. The finite volume groups split further into two categories according to whether the fundamental polygon is compact (after closure in $\hat{\mathbb{C}}$) or not. In the first case we call Γ a *co-compact group*.

Suppose that the polygon $\overline{F}$ is not compact. Then F must have a vertex on $\hat{\mathbb{R}}$, and since the two sides of F which meet at such a vertex are tangent (because they are orthogonal to $\hat{\mathbb{R}}$), they form a *cusp*.

Proposition 2.4. *A fundamental domain of a finite volume group can be chosen as a polygon all of whose cuspidal vertices are inequivalent.*

For a fundamental domain of Γ chosen as in Proposition 2.4, the two sides joined at a cuspidal vertex are equivalent; so the side-pairing motion fixes the vertex, is a parabolic motion, and generates the stability group of the vertex. For this reason a cuspidal vertex is also called a *parabolic vertex*. Conversely, cusps for Γ are exactly the fixed points of parabolic motions of Γ. Hence, we have

Proposition 2.5. *A finite volume group is co-compact if and only if it has no parabolic elements.*

A fundamental polygon all of whose cuspidal vertices are distinct mod Γ will be convenient for various computations. Throughout, we denote cusps by gothic characters $\mathfrak{a}$, $\mathfrak{b}$, $\mathfrak{c}, \ldots$. The stability group of a cusp $\mathfrak{a}$ is an infinite cyclic group generated by a parabolic motion,

$$\Gamma_{\mathfrak{a}} = \big\{ \gamma \in \Gamma : \gamma \mathfrak{a} = \mathfrak{a} \big\} = \langle \gamma_{\mathfrak{a}} \rangle \,,$$

say. Recall we are talking of subgroups of $PSL_2(\mathbb{R})$ rather then of $SL_2(\mathbb{R})$. There exists $\sigma_\mathfrak{a} \in G$ such that

$$(2.1) \qquad\qquad \sigma_\mathfrak{a} \infty = \mathfrak{a}, \qquad \sigma_\mathfrak{a}^{-1} \gamma_\mathfrak{a} \sigma_\mathfrak{a} = \begin{pmatrix} 1 & 1 \\ & 1 \end{pmatrix}.$$

We shall call $\sigma_\mathfrak{a}$ a scaling matrix of the cusp $\mathfrak{a}$; it is determined up to composition with a translation from the right side. The semi-strip

$$(2.2) \qquad\qquad P(Y) = \big\{ z = x + iy : \ 0 < x < 1, \ y \ge Y \big\}$$

is mapped into the cuspidal zone

$$(2.3) \qquad\qquad F_\mathfrak{a}(Y) = \sigma_\mathfrak{a} P(Y).$$

For Y sufficiently large the cuspidal zones are disjoint, and the set

$$F(Y) = F \setminus \bigcup_\mathfrak{a} F_\mathfrak{a}(Y)$$

is compact (after closure) and adjacent to each $F_\mathfrak{a}(Y)$ along the horocycles

$$(2.4) \qquad \sigma_\mathfrak{a} L(Y), \qquad L(Y) = \big\{ z = x + iY : \ 0 < x < 1 \big\}.$$

In this way the fundamental polygon F is partitioned into the central part $F(Y)$ and the cuspidal zones $F_\mathfrak{a}(Y)$ so that

$$(2.5) \qquad\qquad F = F(Y) \cup \bigcup_\mathfrak{a} F_\mathfrak{a}(Y).$$

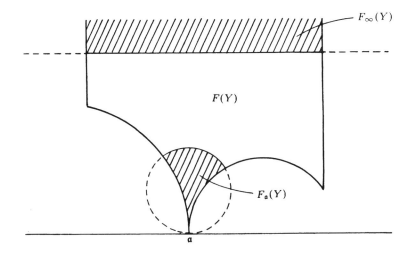

Figure 7. Cuspidal zones and the central part.

Let Γ be a finite volume group. The quotient space $\Gamma\backslash\mathbb{H}$ (the set of orbits) is equipped with the topology in which the natural projection $\pi :$ $\mathbb{H} \longrightarrow \Gamma\backslash\mathbb{H}$ is continuous. In fact, $\Gamma\backslash\mathbb{H}$ is a Hausdorff connected space, and with properly chosen analytic charts it becomes a Riemann surface. If the group Γ contains only hyperbolic elements, and the identity, then $\Gamma\backslash\mathbb{H}$ is a compact, smooth surface of genus $g \geq 2$. If Γ has elliptic elements, then $\Gamma\backslash\mathbb{H}$ has branch points at the fixed points of the elliptic motions. If Γ has parabolic elements, then $\Gamma\backslash\mathbb{H}$ is not compact; in this case one usually compactifies $\Gamma\backslash\mathbb{H}$ by adding cusps with suitable charts.

It is easy to think of the Riemann surface $\Gamma\backslash\mathbb{H}$ as being constructed from a normal polygon by glueing pairs of congruent sides at equivalent points.

There exists a quite explicit construction (in terms of matrix entries of the group elements) of a fundamental domain that is more practical for us than the normal polygon. Suppose Γ is not co-compact. By conjugation we may require that $\mathfrak{a} = \infty$ is a cusp whose stability group Γ_∞ is generated by $\begin{pmatrix} 1 & 1 \\ & 1 \end{pmatrix}$, so a fundamental domain of Γ_∞ is any vertical strip of width 1. We choose

$$F_\infty = \big\{ z \in \mathbb{H} : \ \beta < x < \beta + 1 \big\}.$$

Define F to be the subset of F_∞ which is exterior to all the isometric circles C_γ with $\gamma \in \Gamma$, $\gamma \notin \Gamma_\infty$ (see Section 1.1), *i.e.*

$$(2.6) \qquad F = \big\{ z \in F_\infty : \ \mathrm{Im}\, z > \mathrm{Im}\, \gamma z \ \text{ for all } \gamma \in \Gamma, \gamma \notin \Gamma_\infty \big\}.$$

Thus, F consists of points of deformation less than 1 inside the strip F_∞. One can show that the polygon (2.6) is a fundamental domain of Γ. We shall call it the *standard polygon* (it was first introduced by L. R. Ford [For]). This polygon will be used effectively in Section 2.6 to establish various estimates which are uniform with respect to the group.

2.3. Basic examples

There are various ways to construct a finite volume group. One may start by drawing a convex hyperbolic polygon $F \subset \mathbb{H}$ of an even number of sides and finite volume. However, not every such F is a fundamental domain of a group $\Gamma \subset PSL_2(\mathbb{R})$. The polygon F must satisfy various conditions. For example, since the action of Γ on F tessellates $\mathbb{H}$, the sum of interior angles of F at equivalent vertices is of type $2\pi m^{-1}$, where m is the order of the stability groups for these vertices. Poincaré has given a complete characterization of fundamental polygons of discrete groups which is quite appealing (the angle conditions are barely insufficient).

A subgroup $\Gamma \subset PSL_2(\mathbb{R})$ is called a *triangular group of type* (α, β, γ) if it is generated by the reflections on the sides of some triangle with interior angles α, β, γ (note that it always takes an even number of reflections to make a group element, an analytic automorphism of $\hat{\mathbb{C}}$). Since triangles with the same angles are congruent, all groups of the same type (regardless of the ordering of angles) are conjugate in $PSL_2(\mathbb{R})$. A triangular group is discrete if and only if it is of type $(\pi/p, \pi/q, \pi/r)$, where p, q, r are integers with

$$2 \leq p, q, r \leq +\infty, \qquad 0 < \frac{1}{p} + \frac{1}{q} + \frac{1}{r} < 1.$$

An example of a triangular group is the *Hecke group* Γ_q which is generated by the involution $z \mapsto -1/z$ and the translation $z \mapsto z + 2\cos(\pi/q)$, where q is an integer greater than 2. Therefore, a fundamental domain of Γ_q is the triangle $F = \{z \in \mathbb{H} : |x| < \lambda/2, |z| > 1\}$ with $\lambda = 2\cos(\pi/q)$ of volume $|F| = \pi(1 - 2/q)$. Moreover, i is an elliptic vertex of order 2, $\zeta_q = e(1/2q)$ is an elliptic vertex of order q, ∞ is the cusp, and $g = 0$ is the genus of $\Gamma_q \backslash \mathbb{H}$. One can show that Γ_q is maximal, *i.e.* is not contained in any smaller volume group.

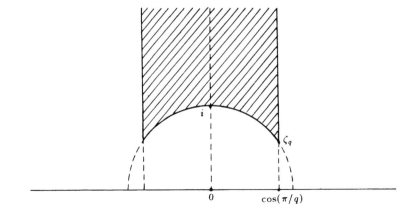

Figure 8.   Fundamental domain for Γ_q.

There is a general group-theoretical recipe for finite volume groups, but it does not reveal much geometry (see R. Fricke and F. Klein [Fr-Kl]).

Proposition 2.6. *Any finite volume subgroup of $PSL_2(\mathbb{R})$ is generated by primitive motions $A_1, \ldots, A_g, B_1, \ldots, B_g, E_1, \ldots, E_\ell, P_1, \ldots, P_h$ satisfying the relations*

$$[A_1, B_1] \cdots [A_g, B_g] E_1 \cdots E_\ell P_1 \cdots P_h = 1, \qquad E_j^{m_j} = 1,$$

where A_j, B_j are hyperbolic motions, $[A_j, B_j]$ stands for the commutator, g is the genus of $\Gamma \backslash \mathbb{H}$, E_j are elliptic motions of order $m_j \geq 2$, P_j are parabolic motions and h is the number of inequivalent cusps.

The symbol $(g; m_1, \ldots, m_\ell; h)$ is group invariant, and is called the *signature* of Γ; it satisfies the Gauss-Bonnet formula

$$(2.7) \qquad 2g - 2 + \sum_{j=1}^{\ell} \left(1 - \frac{1}{m_j}\right) + h = \frac{|F|}{2\pi} \, .$$

Poincaré showed that the positivity of the left side of (2.7) guarantees the existence of Γ with the given signature.

Of all the finite volume groups, the most attractive ones for number theory are the arithmetic groups. Since any comprehensive definition is rather involved (*cf.* [Kat]), we content ourselves with basic examples.

First we introduce the following group (a special type of quaternion group)

$$(2.8) \qquad \Gamma(n, p) = \left\{ \begin{pmatrix} a + b\sqrt{n} & (c + d\sqrt{n})\sqrt{p} \\ (c - d\sqrt{n})\sqrt{p} & a - b\sqrt{n} \end{pmatrix} : \right.$$
$$\left. a, b, c, d \in \mathbb{Z}, \ a^2 - b^2 n - c^2 p + d^2 np = 1 \right\} \, .$$

Here n is a positive integer, and $p \equiv 1 \pmod 4$ is a prime number such that $(n/p) = -1$, *i.e.* n and $-n$ are not quadratic residues modulo p. Using this property, one can show that every element different from ± 1 has trace of absolute value greater than 2; hence it is hyperbolic. Therefore $\Gamma(n, p)$ is discrete by a general result of Nielsen (see Section 2.1) and co-compact by Proposition 2.5.

Our next example is the familiar *modular group*

$$(2.9) \qquad SL_2(\mathbb{Z}) = \left\{ \begin{pmatrix} a & b \\ c & d \end{pmatrix} : a, b, c, d \in \mathbb{Z}, \ ad - bc = 1 \right\}$$

with its fundamental domain $F = \{z = x + iy : |x| < 1/2, \ |z| > 1\}$ which is the normal polygon $D(iv)$ with $v > 1$ as well as a standard polygon. Moreover, i is an elliptic vertex of order 2, $\zeta = (1 + i\sqrt{3})/2$ is an elliptic vertex of order 3, ∞ is the only cusp (up to equivalence) and the genus of $SL_2(\mathbb{Z}) \backslash \mathbb{H}$ is $g = 0$.

Let N be a positive integer. The *principal congruence group of level N*, denoted by $\Gamma(N)$, is the subgroup of the modular group consisting of matrices congruent to the identity modulo N, *i.e.*

$$(2.10) \qquad \Gamma(N) = \left\{ \gamma \in SL_2(\mathbb{Z}) : \gamma \equiv \begin{pmatrix} 1 & \\ & 1 \end{pmatrix} \pmod N \right\} \, .$$

$\Gamma(N)$ is a normal subgroup of $\Gamma(1) = SL_2(\mathbb{Z})$ of index

$$\mu = [\Gamma(1) : \Gamma(N)] = N^3 \prod_{p|N}(1 - p^{-2}).$$

The number of inequivalent cusps is

$$h = \mu N^{-1} = N^2 \prod_{p|N}(1 - p^{-2}).$$

All cusps for $\Gamma(N)$ are rational points $\mathfrak{a} = u/v$ with $(u, v) = 1$ (under the convention that $\pm 1/0 = \infty$). Two cusps $\mathfrak{a} = u/v$ and $\mathfrak{a}' = u'/v'$ are equivalent if and only if $\pm\begin{bmatrix} u \\ v \end{bmatrix} \equiv \begin{bmatrix} u' \\ v' \end{bmatrix}$ (mod N). There are no elliptic elements in $\Gamma(N)$, if $N > 3$.

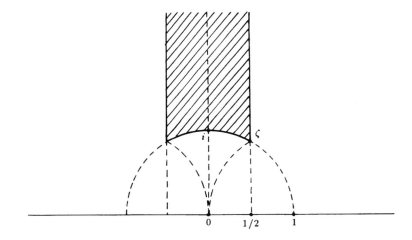

Figure 9.   A fundamental domain for the modular group.

Any subgroup of the modular group containing $\Gamma(N)$ is called a *congruence group of level N*. Two basic examples are

$$\Gamma_0(N) = \left\{ \gamma \in SL_2(\mathbb{Z}) : \ \gamma \equiv \begin{pmatrix} * & * \\ & * \end{pmatrix} \ (\text{mod } N) \right\},$$

$$\Gamma_1(N) = \left\{ \gamma \in SL_2(\mathbb{Z}) : \ \gamma \equiv \begin{pmatrix} 1 & * \\ & 1 \end{pmatrix} \ (\text{mod } N) \right\}.$$

The group $\Gamma_0(N)$ is the most popular; it is called the *Hecke congruence group of level N*, and has index

$$(2.11) \qquad\qquad \mu = [\Gamma_0(1) : \Gamma_0(N)] = N \prod_{p|N}\left(1 + \frac{1}{p}\right).$$

The number of inequivalent elliptic fixed points of order 2 is

(2.12)
$$\nu_2 = \prod_{p|N} \left(1 + \left(\frac{-1}{p}\right)\right) \quad \text{if } 4 \nmid N \,,$$

and the number of those of order 3 is

(2.13)
$$\nu_3 = \prod_{p|N} \left(1 + \left(\frac{-3}{p}\right)\right) \quad \text{if } 9 \nmid N \,.$$

There are no elliptic fixed points of corresponding order if $4|N$ or $9|N$, respectively. Every cusp for $\Gamma_0(N)$ is equivalent to a rational point $\mathfrak{a} = u/v$ with $v \geq 1$, $v|N$, $(u, v) = 1$. Two cusps $\mathfrak{a} = u/v$, $\mathfrak{a}' = u'/v'$ of such form are equivalent if and only if $v = v'$ and $u \equiv u' \pmod{(v, N/v)}$. Therefore the number of inequivalent cusps for $\Gamma_0(N)$ is given by

(2.14)
$$h = \sum_{vw=N} \varphi((v, w)) \,.$$

In particular, if N is prime there are two inequivalent cusps for $\Gamma_0(N)$ at ∞ and 0; they are equivalent to $1/N$ and 1, respectively. All of the above properties of congruence groups can be found in [Sh].

2.4. The double coset decomposition.

Let Γ be a group of finite volume but not co-compact. For such a group the Fourier expansion at cusps is available to help examine automorphic forms through the coefficients. The Fourier expansion will be derived from a decomposition of Γ into double cosets with respect to the stability groups of cusps. Choose two cusps $\mathfrak{a}$, $\mathfrak{b}$ for Γ (not necessarily distinct) and the corresponding scaling matrices $\sigma_{\mathfrak{a}}$, $\sigma_{\mathfrak{b}}$ (determined up to translations from the right side, see (2.1)). Let us recall that $\sigma_{\mathfrak{a}}\infty = \mathfrak{a}$, $\sigma_{\mathfrak{a}}^{-1}\Gamma_{\mathfrak{a}}\sigma_{\mathfrak{a}} = B$ and $\sigma_{\mathfrak{b}}\infty = \mathfrak{b}$, $\sigma_{\mathfrak{b}}^{-1}\Gamma_{\mathfrak{b}}\sigma_{\mathfrak{b}} = B$, where $\Gamma_{\mathfrak{a}}$, $\Gamma_{\mathfrak{b}}$ are the stability groups of cusps and B denotes the group of integral translations, *i.e.*

(2.15)
$$B = \left\{ \begin{pmatrix} 1 & b \\ & 1 \end{pmatrix} : b \in \mathbb{Z} \right\} \,.$$

We shall partition the set $\sigma_{\mathfrak{a}}^{-1}\Gamma\sigma_{\mathfrak{b}}$ into double cosets with respect to B.

First let us examine the subset of the upper-triangular matrices, *i.e.* those having the fixed point at ∞,

$$\Omega_\infty = \left\{ \begin{pmatrix} * & * \\ & * \end{pmatrix} \in \sigma_{\mathfrak{a}}^{-1}\Gamma\sigma_{\mathfrak{b}} \right\} \,.$$

Suppose Ω_∞ is not empty. Take $\omega_\infty = \sigma_\mathfrak{a}^{-1} \gamma \sigma_\mathfrak{b} \in \Omega_\infty$ with $\gamma \in \Gamma$ and evaluate at $\mathfrak{b}$, getting $\gamma \mathfrak{b} = \sigma_\mathfrak{a} \omega_\infty \infty = \sigma_\mathfrak{a} \infty = \mathfrak{a}$. This shows that the cusps $\mathfrak{a}$, $\mathfrak{b}$ are equivalent, the stability groups are conjugate, and $\omega_\infty = \sigma_\mathfrak{a}^{-1} \gamma \sigma_\mathfrak{b}$ is a translation. Now suppose $\omega_1 = \sigma_\mathfrak{a}^{-1} \gamma_1 \sigma_\mathfrak{b}$ is another element of Ω_∞. We obtain

$$\gamma \gamma_1^{-1} \mathfrak{a} = \sigma_\mathfrak{a} \omega_\infty \omega_1^{-1} \sigma_\mathfrak{a}^{-1} \mathfrak{a} = \sigma_\mathfrak{a} \omega_\infty \omega_1^{-1} \infty = \sigma_\mathfrak{a} \infty = \mathfrak{a},$$

showing that $\gamma \gamma_1^{-1} \in \Gamma_\mathfrak{a}$; hence $\omega_\infty \omega_1^{-1} = \sigma_\mathfrak{a}^{-1} \gamma \gamma_1^{-1} \sigma_\mathfrak{a} \in \sigma_\mathfrak{a}^{-1} \Gamma_\mathfrak{a} \sigma_\mathfrak{a} = B$. Combining the results, we conclude that the subset Ω_∞ is not empty if and only if the cusps $\mathfrak{a}$, $\mathfrak{b}$ are equivalent, in which case

$$(2.16) \qquad \Omega_\infty = B \omega_\infty B = \omega_\infty B = B \omega_\infty$$

for some

$$(2.17) \qquad \omega_\infty = \begin{pmatrix} 1 & * \\ & 1 \end{pmatrix} \in \sigma_\mathfrak{a}^{-1} \Gamma \sigma_\mathfrak{b} .$$

All other elements of $\sigma_\mathfrak{a}^{-1} \Gamma \sigma_\mathfrak{b}$ fall into one of the double cosets

$$(2.18) \qquad \Omega_{d/c} = B \omega_{d/c} B$$

for some

$$(2.19) \qquad \omega_{d/c} = \begin{pmatrix} * & * \\ c & d \end{pmatrix} \in \sigma_\mathfrak{a}^{-1} \Gamma \sigma_\mathfrak{b} ,$$

with $c > 0$. The relation

$$(2.20) \qquad \begin{pmatrix} 1 & m \\ & 1 \end{pmatrix} \begin{pmatrix} a & * \\ c & d \end{pmatrix} \begin{pmatrix} 1 & n \\ & 1 \end{pmatrix} = \begin{pmatrix} a + cm & * \\ c & d + cn \end{pmatrix}$$

shows that the double coset $\Omega_{d/c}$ determines c uniquely, whereas a and d are determined modulo integral multiples of c. In fact, $\Omega_{d/c}$ does not depend on the upper row of $\omega_{d/c}$. To see this, take $\omega = \begin{pmatrix} a & * \\ c & d \end{pmatrix}$, $\omega_1 = \begin{pmatrix} a_1 & * \\ c & d \end{pmatrix}$, two elements of $\sigma_\mathfrak{a}^{-1} \Gamma \sigma_\mathfrak{b}$ with the same lower row. Setting $\gamma = \sigma_\mathfrak{a} \omega \sigma_\mathfrak{b}^{-1} \in \Gamma$, $\gamma_1 = \sigma_\mathfrak{a} \omega_1 \sigma_\mathfrak{b}^{-1} \in \Gamma$, we obtain $\gamma \gamma_1^{-1} = \sigma_\mathfrak{a} \omega \omega_1^{-1} \sigma_\mathfrak{a}^{-1} = \sigma_\mathfrak{a} \begin{pmatrix} * & * \\ & * \end{pmatrix} \sigma_\mathfrak{a}^{-1}$. Evaluating at $\mathfrak{a}$, we infer that $\gamma \gamma_1^{-1} \in \Gamma_\mathfrak{a}$, whence $\omega \omega_1^{-1} \in B$, which shows that $a \equiv a_1 \pmod{c}$, as claimed.

By the above investigation we have established the following fact.

Theorem 2.7.. *Let $\mathfrak{a}$, $\mathfrak{b}$ be cusps for Γ. We then have a disjoint union*

$$(2.21) \qquad \sigma_\mathfrak{a}^{-1} \Gamma \sigma_\mathfrak{b} = \delta_{\mathfrak{a}\mathfrak{b}} \, \Omega_\infty \cup \bigcup_{c > 0} \bigcup_{d (\bmod c)} \Omega_{d/c},$$

where $\delta_{\mathfrak{a}\mathfrak{b}} = 1$ if $\mathfrak{a}$, $\mathfrak{b}$ are equivalent; otherwise it vanishes. Moreover, c, d run over numbers such that $\sigma_{\mathfrak{a}}^{-1}\Gamma\sigma_{\mathfrak{b}}$ contains $\begin{pmatrix} * & * \\ c & d \end{pmatrix}$.

As an example, take the Hecke congruence group $\Gamma = \Gamma_0(q)$ and the cusps ∞, 0. The scaling matrices are

$$\sigma_\infty = \begin{pmatrix} 1 & \\ & 1 \end{pmatrix}, \qquad \sigma_0 = \begin{pmatrix} & -1/\sqrt{q} \\ \sqrt{q} & \end{pmatrix}.$$

We have

$$\sigma_\infty^{-1}\Gamma\sigma_\infty = \sigma_0^{-1}\Gamma\sigma_0 = \left\{ \begin{pmatrix} \alpha & * \\ \gamma q & \delta \end{pmatrix} : \ \alpha\delta \equiv 1 \ (\mathrm{mod}\ \gamma q) \right\}$$

and

$$\sigma_\infty^{-1}\Gamma\sigma_0 = \left\{ \begin{pmatrix} \alpha\sqrt{q} & * \\ \gamma\sqrt{q} & \delta\sqrt{q} \end{pmatrix} : \ \alpha\delta q \equiv 1 \ (\mathrm{mod}\ \gamma) \right\}.$$

For a cusp $\mathfrak{a} = 1/v$ with $vw = q$, $(v, w) = 1$ we may take

$$\sigma_{\mathfrak{a}} = \begin{pmatrix} \sqrt{w} & \\ v\sqrt{w} & 1/\sqrt{w} \end{pmatrix}.$$

We find that

$$\sigma_{\mathfrak{a}}^{-1}\Gamma\sigma_{\mathfrak{a}} = \left\{ \begin{pmatrix} \alpha & * \\ \gamma q & \delta \end{pmatrix} : \ \alpha\delta \equiv 1 \ (\mathrm{mod}\ \gamma v), \ (\alpha + \gamma)(\delta - \gamma) \equiv 1 \ (\mathrm{mod}\ w) \right\}$$

and

$$\sigma_\infty^{-1}\Gamma\sigma_{\mathfrak{a}} = \left\{ \begin{pmatrix} \alpha\sqrt{w} & * \\ \gamma v\sqrt{w} & \delta/\sqrt{w} \end{pmatrix} : \ \alpha\delta \equiv 1 \ (\mathrm{mod}\ \gamma v), \ \gamma \equiv \delta (\mathrm{mod}\ w) \right\}.$$

In the above matrices α, δ, γ are integers. For other pairs of cusps the sets $\sigma_{\mathfrak{a}}^{-1}\Gamma\sigma_{\mathfrak{b}}$ have a similar structure.

2.5. Kloosterman sums.

The double coset decomposition (2.21) is a tool for working with the group Γ by means of additive characters. Specifically, to a double coset $\Omega_{d/c} = B\,\omega_{d/c}B$ we shall attach the character

$$\Omega_{d/c}(m) = e\left(m\,\frac{d}{c}\right), \qquad m \in \mathbb{Z}.$$

Another character is attached to the inverse coset $\Omega_{d/c}^{-1} = B\omega_{d/c}^{-1}B = B\omega_{-a/c}B = \Omega_{-a/c}$:

$$\Omega_{-a/c}(n) = e\left(-n\,\frac{a}{c}\right), \qquad n \in \mathbb{Z}.$$

Given $m, n \in \mathbb{Z}$ and any c (which need not be an integer) in the set

$$(2.22) \qquad \mathcal{C}_{\mathfrak{ab}} = \left\{ c > 0 : \begin{pmatrix} * & * \\ c & * \end{pmatrix} \in \sigma_{\mathfrak{a}}^{-1} \Gamma \sigma_{\mathfrak{b}} \right\},$$

the *Kloosterman sum* is created by convolving these characters as follows:

$$(2.23) \qquad \mathcal{S}_{\mathfrak{ab}}(m, n; c) = \sum_{\begin{pmatrix} a & * \\ c & d \end{pmatrix} \in B \backslash \sigma_{\mathfrak{a}}^{-1} \Gamma \sigma_{\mathfrak{b}} / B} e\left(m \frac{d}{c} + n \frac{a}{c} \right).$$

We shall refer to c as the *modulus* and to m, n as *frequencies*. Note the symmetries $\mathcal{C}_{\mathfrak{ab}} = \mathcal{C}_{\mathfrak{ba}}$ and $\mathcal{S}_{\mathfrak{ab}}(m, n; c) = \mathcal{S}_{\mathfrak{ba}}(n, m; c)$. Observe that the Kloosterman sum depends on the choice of the scaling matrices in the following simple fashion (it changes only by fixed additive characters):

$$\mathcal{S}_{\mathfrak{a'b'}}(m, n; c) = e(\alpha m + \beta n)\, \mathcal{S}_{\mathfrak{ab}}(m, n; c),$$

if $\mathfrak{a'} = \tau_{\mathfrak{a}} \mathfrak{a}$, $\sigma_{\mathfrak{a'}} = \tau_{\mathfrak{a}} \sigma_{\mathfrak{a}} n(\alpha)$ and $\mathfrak{b'} = \tau_{\mathfrak{b}} \mathfrak{b}$, $\sigma_{\mathfrak{b'}} = \tau_{\mathfrak{b}} \sigma_{\mathfrak{b}} n(\beta)$ for some $\tau_{\mathfrak{a}}, \tau_{\mathfrak{b}}$ in Γ and $n(\alpha), n(\beta)$ in N. In particular, this shows that

$$(2.24) \qquad \mathcal{S}_{\mathfrak{ab}}(0, 0; c) = \#\left\{ d\,(\mathrm{mod}\ c) : \begin{pmatrix} * & * \\ c & d \end{pmatrix} \in \sigma_{\mathfrak{a}}^{-1} \Gamma \sigma_{\mathfrak{b}} \right\}$$

depends only on the equivalence classes of cusps, but not on their representatives nor on the choice of scaling matrices.

Let us see closely what the above constructions yield for the modular group $\Gamma = SL_2(\mathbb{Z})$. In this case there is only one cusp $\mathfrak{a} = \mathfrak{b} = \infty$, for which we obtain the classical Kloosterman sum

$$\mathcal{S}(m, n; c) = \sum_{ad \equiv 1\,(\mathrm{mod}\ c)} e\left(\frac{dm + an}{c} \right)$$

defined for all positive integer moduli c. We have a deep bound:

$$(2.25) \qquad |\mathcal{S}(m, n; c)| \leq (m, n, c)^{1/2} c^{1/2} \tau(c),$$

where $\tau(c)$ is the divisor function, $\tau(c) \ll c^{\varepsilon}$. For c prime (the hardest case) this bound was derived by A. Weil [We] as a consequence of the Riemann hypothesis for curves over finite fields. The special case $n = 0$ is simple; the Kloosterman sum reduces to the *Ramanujan sum*

$$(2.26) \qquad \mathcal{S}(m, 0; c) = \sum_{d\,(\mathrm{mod}\ c)}^{*} e\left(\frac{dm}{c} \right) = \sum_{\delta | (c, m)} \mu\left(\frac{c}{\delta} \right) \delta,$$

where the star restricts the summation to the classes prime to the modulus; hence the generating Dirichlet series for Ramanujan sums is equal to

$$(2.27) \qquad \sum_{c=1}^{\infty} c^{-2s} S(m, 0; c) = \zeta(2s)^{-1} \sum_{\delta | m} \delta^{1-2s} \, .$$

For $m = n = 0$ we get the Euler function $S(0, 0; c) = \varphi(c)$ and

$$(2.28) \qquad \sum_{c=1}^{\infty} c^{-2s} S(0, 0; c) = \frac{\zeta(2s-1)}{\zeta(2s)} \, .$$

2.6. Basic estimates.

Let us return to the general case of a finite volume group Γ which is not co-compact. In applications we shall need some control over the number of cosets in the decomposition (2.21). Let $c(\mathfrak{a}, \mathfrak{b})$ denote the smallest element of the set $\mathcal{C}_{\mathfrak{a}\mathfrak{b}}$. Put $c_{\mathfrak{a}} = c(\mathfrak{a}, \mathfrak{a})$, *i.e.*

$$(2.30) \qquad c_{\mathfrak{a}} = \min \left\{ c > 0 : \begin{pmatrix} * & * \\ c & * \end{pmatrix} \in \sigma_{\mathfrak{a}}^{-1} \Gamma \sigma_{\mathfrak{a}} \right\} .$$

That $c_{\mathfrak{a}}$ exists is seen from the construction of the standard polygon for the group $\sigma_{\mathfrak{a}}^{-1} \Gamma \sigma_{\mathfrak{a}}$; $c_{\mathfrak{a}}^{-1}$ is the radius of the largest isometric circle. Since the polygon contains the semi-strip $P(c_{\mathfrak{a}}^{-1})$ of volume $c_{\mathfrak{a}}$, we have

$$(2.31) \qquad c_{\mathfrak{a}} < |F| \, .$$

First, for any c in the set $\mathcal{C}_{\mathfrak{a}\mathfrak{b}}$ we estimate the number (2.24). Surprisingly, there is no sharp bound available for each $\mathcal{S}_{\mathfrak{a}\mathfrak{b}}(0, 0; c)$ individually. We show the following:

Proposition 2.8. *For any $c \in \mathcal{C}_{\mathfrak{a}\mathfrak{b}}$ we have*

$$(2.32) \qquad \mathcal{S}_{\mathfrak{a}\mathfrak{b}}(0, 0; c) \leq c_{\mathfrak{a}\mathfrak{b}}^{-1} c^2,$$

where $c_{\mathfrak{a}\mathfrak{b}} = \max\{c_{\mathfrak{a}}, c_{\mathfrak{b}}\}$, and we have superior bound on average:

$$(2.33) \qquad \sum_{c \leq X} c^{-1} \mathcal{S}_{\mathfrak{a}\mathfrak{b}}(0, 0; c) \leq c_{\mathfrak{a}\mathfrak{b}}^{-1} X \, .$$

Proof. By symmetry we can assume without loss of generality that $c_{\mathfrak{a}} \geq c_{\mathfrak{b}}$. If $\gamma = \begin{pmatrix} * & * \\ c & d \end{pmatrix}$ and $\gamma' = \begin{pmatrix} * & * \\ c' & d' \end{pmatrix}$, where $c, c' > 0$, are both in $\sigma_{\mathfrak{a}}^{-1} \Gamma \sigma_{\mathfrak{b}}$, then

$$\gamma'' = \gamma' \gamma^{-1} = \begin{pmatrix} * & * \\ c'' & * \end{pmatrix} \in \sigma_{\mathfrak{a}}^{-1} \Gamma \sigma_{\mathfrak{a}},$$

where $c'' = c'd - cd'$. If $c'' = 0$, then the cusps $\mathfrak{a}$, $\mathfrak{b}$ are equivalent, $\gamma'' = \begin{pmatrix} 1 & * \\ & 1 \end{pmatrix}$, $c' = c$ and $d' = d$. If $c'' \neq 0$, then $|c''| \geq c_{\mathfrak{a}}$, whence

$$(2.34) \qquad \left| \frac{d'}{c'} - \frac{d}{c} \right| \geq \frac{c_{\mathfrak{a}}}{c\,c'} \, .$$

In particular, for $c' = c$ this yields

$$(2.35) \qquad |d' - d| \geq c_{\mathfrak{a}} \, c^{-1} \, .$$

Hence, the bound (2.32) is derived by applying the box principle. Similarly, if $0 < c, c' \leq X$, we get from (2.34) that

$$(2.36) \qquad \left| \frac{d'}{c'} - \frac{d}{c} \right| \geq c_{\mathfrak{a}} \, c^{-1} X^{-1} \, .$$

Summing this inequality over $c \leq X$ and $0 \leq d < c$, where d'/c' is chosen to be the succesive point to d/c, we get (2.33).

Notice that $c(\mathfrak{a}, \mathfrak{b})^2 \geq c_{\mathfrak{a}\mathfrak{b}}$, which one can deduce by applying $\mathcal{S}_{\mathfrak{a}\mathfrak{b}}(0, 0; c) \geq 1$ to (2.32) with $c = c(\mathfrak{a}, \mathfrak{b})$. Incidentally, (2.32) follows from (2.33).

Corollary 2.9. *The Kloosterman sums satisfy the following trivial bounds:*

$$(2.37) \qquad |\mathcal{S}_{\mathfrak{a}\mathfrak{b}}(m, n; c)| \leq c_{\mathfrak{a}\mathfrak{b}}^{-1} c^2$$

and

$$(2.38) \qquad \sum_{c \leq X} c^{-1} |\mathcal{S}_{\mathfrak{a}\mathfrak{b}}(m, n; c)| \leq c_{\mathfrak{a}\mathfrak{b}}^{-1} X \, .$$

Lemma 2.10. *Let $\mathfrak{a}$ be a cusp for Γ, $z \in \mathbb{H}$ and $Y > 0$. We have*

$$(2.39) \qquad \#\left\{ \gamma \in \Gamma_{\mathfrak{a}} \backslash \Gamma : \ \mathrm{Im}\, \sigma_{\mathfrak{a}}^{-1} \gamma z > Y \right\} < 1 + \frac{10}{c_{\mathfrak{a}} Y} \, .$$

Proof. Conjugating the group, we can assume that $\mathfrak{a} = \infty$, $\sigma_{\mathfrak{a}} = 1$ and $\Gamma_{\mathfrak{a}} = B$. Then the fundamental domain of $\Gamma_{\mathfrak{a}}$ is the strip

$$P = \{z = x + iy : \ 0 < x < 1, \ y > 0\}\,.$$

Let F be the standard polygon of Γ, so F consists of points in P of deformation less than 1. For the proof we may assume that $z \in F$. Then for any $\gamma = \begin{pmatrix} * & * \\ c & d \end{pmatrix} \in \Gamma$ with $c > 0$ the point γz lies on the isometric circle $|cz+d| = 1$ or in its interior. In any case we have $|cz+d| \geq 1$. Since $\operatorname{Im} \gamma z = y|cz+d|^{-2} > Y$, this implies $y > Y$, $c < y^{-1/2}Y^{-1/2}$, $|cx+d| < y^{1/2}Y^{-1/2}$. By the last inequality and the spacing property (2.34) we estimate the number of pairs $\{c, d\}$ with $C \leq c < 2C$ by $1 + 8\,c_{\mathfrak{a}}^{-1} C\, y^{1/2} Y^{-1/2} \leq 10\, c_{\mathfrak{a}}^{-1} C\, y^{1/2} Y^{-1/2}$. Adding these bounds for $C = 2^{-n} y^{-1/2} Y^{-1/2}$, $n \geq 1$, we get $10\, c_{\mathfrak{a}}^{-1} Y^{-1}$. This is an estimate for the number of relevant γ's not in $\Gamma_{\mathfrak{a}}$. Finally, adding 1 to account for $\Gamma_{\mathfrak{a}}$, we obtain what is claimed.

Lemma 2.11. *Let $\mathfrak{a}$ be a cusp of Γ, $z, w \in \mathbb{H}$ and $\delta > 0$. We have*

$$
\begin{aligned}
(2.40) \qquad & \#\{\gamma \in \sigma_{\mathfrak{a}}^{-1}\Gamma\sigma_{\mathfrak{a}} : \ u(\gamma z, w) < \delta\} \\
& \ll \sqrt{\delta(\delta + 1)}\,(\operatorname{Im} w + c_{\mathfrak{a}}^{-1}) + (\delta + 1)(c_{\mathfrak{a}} \operatorname{Im} w)^{-1} + 1,
\end{aligned}
$$

where the implied constant is absolute.

Proof. Without loss of generality we can assume that $\mathfrak{a} = \infty$, $\sigma_{\mathfrak{a}} = 1$ and $\Gamma_{\mathfrak{a}} = B$. The condition $u(\gamma z, w) < \delta$ is equivalent to

$$(2.41) \qquad |\gamma z - w| < 2\,(\delta \operatorname{Im} w \operatorname{Im} \gamma z)^{1/2}\,.$$

Taking the imaginary part, we infer that $Y_1 < \operatorname{Im} \gamma z < Y_2$, where $Y_1 = \operatorname{Im} w/4(\delta + 1)$ and $Y_2 = 4(\delta + 1)\operatorname{Im} w$. Looking at the real part, we find that the number of elements $\gamma' \in \Gamma_{\mathfrak{a}}$ such that $\gamma'\gamma$ satisfies (2.41) does not exceed $1 + 4\,(\delta \operatorname{Im} w \operatorname{Im} \gamma z)^{1/2}$. Therefore, the total number of γ's satisfying (2.41) does not exceed

$$\sum_{\substack{\gamma \in \Gamma_{\mathfrak{a}}\backslash\Gamma \\ Y_1 < \operatorname{Im} \gamma z < Y_2}} \left(1 + 4\,(\delta \operatorname{Im} w \operatorname{Im} \gamma z)^{1/2}\right).$$

Applying Lemma 2.10, by partial summation, this is bounded by

$$1 + c_{\mathfrak{a}}^{-1} Y_1^{-1} + (\delta \operatorname{Im} w)^{1/2}(Y_2^{1/2} + c_{\mathfrak{a}}^{-1} Y_1^{-1/2}),$$

which yields (2.40).

We shall often require estimates which are uniform as z runs to infinity through cuspidal zones (see (2.3) and Figure 7). To this end we introduce the *invariant height* of z by

$$(2.42) \qquad\qquad y_\Gamma(z) = \max_{\mathfrak{a}} \max_{\gamma \in \Gamma} \{\operatorname{Im} \sigma_\mathfrak{a}^{-1} \gamma z\}.$$

Thus, we have $y_\Gamma(\sigma_\mathfrak{a} z) = y$ if y is sufficiently large, *i.e.* when z approaches the cusp $\mathfrak{a}$. Observe that $y_\Gamma(z)$ is bounded below by a positive constant depending only on Γ, say

$$(2.43) \qquad\qquad y_\Gamma = \min_{z \in \mathbb{H}} y_\Gamma(z) > 0.$$

For example, if $\Gamma = \Gamma_q$ is the Hecke triangle group, then $y_\Gamma = \sin(\pi/q)$ (see Figure 8).

Corollary 2.12. *Let $z \in \mathbb{H}$ and $\delta > 0$. We have*

$$(2.44) \qquad \#\{\gamma \in \Gamma : \; u(\gamma z, z) < \delta\} \ll \sqrt{\delta(\delta + 1)}\, y_\Gamma(z) + \delta + 1,$$

where the implied constant depends on the group alone.

To derive (2.44), let $\mathfrak{a}$ and $\gamma_1 \in \Gamma$ be such that $\operatorname{Im} \sigma_\mathfrak{a}^{-1} \gamma_1 z$ is maximal. Then apply (2.40) for the cusp $\mathfrak{a}$ and both points equal to $\sigma_\mathfrak{a}^{-1} \gamma_1 z$.

Automorphic Forms

3.1. Introduction.

Let Γ be a finite volume group. A function $f : \mathbb{H} \longrightarrow \mathbb{C}$ is said to be *automorphic with respect to* Γ if it satisfies the periodicity condition

$$f(\gamma z) = f(z), \qquad \text{for all } \gamma \in \Gamma.$$

Therefore, f lives on the Riemann surface $\Gamma \backslash \mathbb{H}$. We denote the space of such functions by $\mathcal{A}(\Gamma \backslash \mathbb{H})$.

Some automorphic functions can be constructed by the method of images. Take a function $p(z)$ of sufficiently rapid decay on $\mathbb{H}$. Then

$$f(z) = \sum_{\gamma \in \Gamma} p(\gamma z) \in \mathcal{A}(\Gamma \backslash \mathbb{H}).$$

Very important automorphic functions are given by a series over the cosets of an infinite subgroup of Γ rather than over the whole group. For such construction, of course, the generating function $p(z)$ must be invariant with respect to this subgroup. If Γ is not co-compact, we take cosets with respect to the stability group of a cusp. In this way we obtain the *Poincaré series*

$$(3.1) \qquad E_{\mathfrak{a}}(z|p) = \sum_{\gamma \in \Gamma_{\mathfrak{a}} \backslash \Gamma} p(\sigma_{\mathfrak{a}}^{-1} \gamma z) \in \mathcal{A}(\Gamma \backslash \mathbb{H}),$$

where $p(z)$ is any function on $\mathbb{H}$ which is B-invariant (periodic in x of period 1) and satisfies a suitable growth condition.

From now on we consider a non-co-compact group Γ. Let $\mathfrak{a}$ be a cusp for Γ, and $\sigma_{\mathfrak{a}}$ a scaling matrix. Any $f \in \mathcal{A}(\Gamma \backslash \mathbb{H})$ satisfies the transformation rule

$$f\left(\sigma_{\mathfrak{a}} \begin{pmatrix} 1 & m \\ & 1 \end{pmatrix} z\right) = f(\sigma_{\mathfrak{a}} z),$$

for all $m \in \mathbb{Z}$; therefore, it makes sense to write the Fourier expansion

$$(3.2) \qquad f(\sigma_{\mathfrak{a}} z) = \sum_{n} f_{\mathfrak{a}n}(y)\, e(nx),$$

where the coefficients are given by

$$f_{\mathfrak{a}n}(y) = \int_0^1 f(\sigma_{\mathfrak{a}} z)\, e(-nx)\, dx\,.$$

If f is smooth, then the series (3.2) converges absolutely and uniformly on compacta.

An automorphic function $f \in \mathcal{A}(\Gamma \backslash \mathbb{H})$ which is an eigenfunction of the Laplace operator

$$(\Delta + \lambda)f = 0\,, \qquad \lambda = s(1-s),$$

is called an *automorphic form* (of Maass [Ma]). We denote by $\mathcal{A}_s(\Gamma \backslash \mathbb{H})$ the space of automorphic forms with respect to Γ for the eigenvalue $\lambda = s(1-s)$. Thus $\mathcal{A}_s(\Gamma \backslash \mathbb{H}) = \mathcal{A}_{1-s}(\Gamma \backslash \mathbb{H}) \subset \mathcal{A}(\Gamma \backslash \mathbb{H})$.

For an automorphic form the Fourier expansion (3.2) can be made more explicit. In this case Proposition 1.5 yields

Theorem 3.1. *Any $f \in \mathcal{A}_s(\Gamma \backslash \mathbb{H})$ satisfying the growth condition*

$$(3.3) \qquad f(\sigma_{\mathfrak{a}} z) = o(e^{2\pi y}) \qquad as \ \ y \to +\infty$$

has the expansion

$$(3.4) \qquad f(\sigma_{\mathfrak{a}} z) = f_{\mathfrak{a}}(y) + \sum_{n \neq 0} \hat{f}_{\mathfrak{a}}(n)\, W_s(nz),$$

where the zero-th term $f_{\mathfrak{a}}(y)$ is a linear combination of the functions (1.23), i.e.

$$(3.5) \qquad f_{\mathfrak{a}}(y) = \frac{A}{2}(y^s + y^{1-s}) + \frac{B}{2s-1}(y^s - y^{1-s})\,.$$

The non-zero coefficients in (3.4) are bounded by

$$(3.6) \qquad \hat{f}_{\mathfrak{a}}(n) \ll e^{\varepsilon |n|}\,,$$

for any $\varepsilon > 0$, with the implied constant depending on ε and f. Therefore, as $y \to +\infty$, we have

$$(3.7) \qquad f(\sigma_{\mathfrak{a}} z) = f_{\mathfrak{a}}(y) + O(e^{-2\pi y})\,.$$

Our objective is to expand an automorphic function into automorphic forms subject to suitable growth conditions. The main results hold in the Hilbert space

$$\mathfrak{L}(\Gamma \backslash \mathbb{H}) = \{f \in \mathcal{A}(\Gamma \backslash \mathbb{H}) : \ \|f\| < \infty\}$$

with the inner product

$$\langle f, g \rangle = \int_F f(z) \, \overline{g}(z) \, d\mu z \, .$$

Observe that bounded automorphic functions are in $\mathfrak{L}(\Gamma \backslash \mathbb{H})$ because F has finite volume. The inner product is a powerful tool for analytic as well as arithmetic studies of automorphic forms. Be aware of the positivity of the norm $\|f\| = \langle f, f \rangle^{1/2}$. Contemplate the arguments exploiting this obvious fact throughtout the lectures.

When f is square integrable we shall improve the estimates of Theorem 3.1 substantially.

Theorem 3.2. *If $f \in \mathcal{A}_s(\Gamma \backslash \mathbb{H}) \cap \mathfrak{L}(\Gamma \backslash \mathbb{H})$, then it has bounded Fourier coefficients. In fact, assuming $\|f\| = 1$ and $\mathrm{Re}\, s \geq 1/2$ (normalization conditions), we have*

(3.8)
$$\sum_{|n| \leq N} |n| \, |\hat{f}_{\mathfrak{a}}(n)|^2 \ll (c_{\mathfrak{a}}^{-1} N + |s|) \, e^{\pi |s|}.$$

Hence

$$f(\sigma_{\mathfrak{a}} z) = f_{\mathfrak{a}}(y) + O\big((\frac{|s|}{y}(1 + \frac{1}{c_{\mathfrak{a}} y}))^{1/2}\big) \, .$$

The zero-th term exists only if $1/2 < s \leq 1$, and it takes the form $f_{\mathfrak{a}}(y) = \hat{f}_{\mathfrak{a}}(0) \, y^{1-s}$ with the coefficient satisfying

$$\hat{f}_{\mathfrak{a}}(0) \ll \big(s - \frac{1}{2}\big)^{1/2} c_{\mathfrak{a}}^{-s+1/2} \, .$$

All the implied constants are absolute.

Proof. We get by Parseval's identity that

$$|f_{\mathfrak{a}}(y)|^2 + \sum_{n \neq 0} |\hat{f}_{\mathfrak{a}}(n) \, W_s(iny)|^2 = \int_0^1 |f(\sigma_{\mathfrak{a}}(x + iy))|^2 \, dx \, .$$

Hence,

$$|\hat{f}_{\mathfrak{a}}(0)|^2 \frac{Y^{1-2s}}{2s - 1} + \sum_{n \neq 0} |\hat{f}_{\mathfrak{a}}(n)|^2 \int_Y^{+\infty} W_s^2(iny) \, y^{-2} \, dy$$

$$= \int_Y^{+\infty} \int_0^1 |f(\sigma_{\mathfrak{a}} z)|^2 \, d\mu z \leq 1 + \frac{10}{c_{\mathfrak{a}} Y},$$

because every orbit $\{\sigma_{\mathfrak{a}}\gamma z : \ \gamma \in \Gamma\}$ has no more than the above number of points in the strip $P(Y)$ by Lemma 2.10. On the other hand, it follows from

$$\int_{|s|/2}^{+\infty} K_{s-1/2}^2(y)\, y^{-1}\, dy \gg |s|^{-1}\, e^{-\pi|s|}$$

that

$$\int_Y^{+\infty} W_s^2(iny)\, y^{-2}\, dy \gg |n|\, |s|^{-1}\, e^{-\pi|s|}\,,$$

if $4\pi|n|Y \le |s|$. Setting $4\pi NY = |s|$, one infers the desired estimate for the sum of non-zero Fourier coefficients. To estimate the zero-th coefficient, take $Y = c_{\mathfrak{a}}^{-1}$.

To estimate $f(\sigma_{\mathfrak{a}}z)$, apply Cauchy's inequality to the Fourier expansion as follows:

$$\begin{aligned}
|f(\sigma_{\mathfrak{a}}z) &- f_{\mathfrak{a}}(y)|^2 \\
&\le \sum_{n\neq 0} |n|\,(|n|+Y)^{-2}\,|\hat{f}_{\mathfrak{a}}(n)|^2 \sum_{n\neq 0} |n|^{-1}(|n|+Y)^2\,|W_s(nz)|^2 \\
&\ll Y^{-2}(c_{\mathfrak{a}}^{-1}Y + |s|)\,|s|\,y^{-1}\,(y^{-1}|s| + Y)^2\,,
\end{aligned}$$

which yields our claim upon taking $Y = |s|/4\pi y$.

3.2. The Eisenstein series.

We begin with the Poincaré series (3.1) evolved from

$$p(z) = \psi(y)\, e(mz),$$

where m is a non-negative integer and ψ is a smooth function on $\mathbb{R}^+$. For the absolute convergence it is sufficient that (see (2.39))

$$(3.9) \qquad\qquad \psi(y) \ll y\,(\log y)^{-2}\,, \qquad \text{as } \ y \to 0\,.$$

This yields a kind of weighted Poincaré series, which was considered by A. Selberg [Se1],

$$E_{\mathfrak{a}m}(z|\psi) = \sum_{\gamma\in\Gamma_{\mathfrak{a}}\backslash\Gamma} \psi(\operatorname{Im}\sigma_{\mathfrak{a}}^{-1}\gamma z)\, e(m\sigma_{\mathfrak{a}}^{-1}\gamma z)\,.$$

For $m = 0$ this becomes a kind of weighted Eisenstein series,

$$(3.10) \qquad\qquad E_{\mathfrak{a}}(z|\psi) = \sum_{\gamma\in\Gamma_{\mathfrak{a}}\backslash\Gamma} \psi(\operatorname{Im}\sigma_{\mathfrak{a}}^{-1}\gamma z)\,.$$

For $\psi(y) = y^s$ with $\operatorname{Re} s > 1$ we obtain the Eisenstein series

$$(3.11) \qquad E_{\mathfrak{a}}(z, s) = \sum_{\gamma \in \Gamma_{\mathfrak{a}} \backslash \Gamma} (\operatorname{Im} \sigma_{\mathfrak{a}}^{-1} \gamma z)^s \,.$$

Since $p(z) = y^s$ is an eigenfunction of Δ with eigenvalue $\lambda = s(1 - s)$, so is the Eisenstein series, *i.e.* $E_{\mathfrak{a}}(z, s) \in \mathcal{A}_s(\Gamma \backslash \mathbb{H})$ if $\operatorname{Re} s > 1$. But $E_{\mathfrak{a}}(z, s)$ is not square integrable over F.

If ψ is compactly supported on $\mathbb{R}^+$, we call $E_{\mathfrak{a}}(z|\psi)$ an *incomplete Eisenstein series*. In this case $E_{\mathfrak{a}}(z|\psi)$ is a bounded automorphic function on $\mathbb{H}$ and hence clearly square integrable over F. The incomplete Eisenstein series is not an automorphic form, because it fails to be an eigenfunction of Δ. However, by Mellin's inversion one can represent the incomplete Eisenstein series as a contour integral of the Eisenstein series:

$$(3.12) \qquad E_{\mathfrak{a}}(z|\psi) = \frac{1}{2\pi i} \int_{(\sigma)} E_{\mathfrak{a}}(z, s) \, \hat{\psi}(s) \, ds \,,$$

where $\sigma > 1$ and

$$(3.13) \qquad \hat{\psi}(s) = \int_0^{+\infty} \psi(y) \, y^{-s-1} \, dy \,.$$

Note that $\hat{\psi}(s) \ll (|s|+1)^{-A}$ (by repeated partial integration) in the vertical strips $\sigma_1 \le \operatorname{Re} s \le \sigma_2$, where σ_1, σ_2 and A are any constants. Hence, it is clear that the integral (3.12) converges absolutely.

To pursue the above analysis we select two linear subspaces: the space $\mathcal{B}(\Gamma \backslash \mathbb{H})$ of smooth and bounded automorphic functions, and the space $\mathcal{E}(\Gamma \backslash \mathbb{H})$ of incomplete Eisenstein series. We have the inclusions $\mathcal{E}(\Gamma \backslash \mathbb{H}) \subset \mathcal{B}(\Gamma \backslash \mathbb{H}) \subset \mathfrak{L}(\Gamma \backslash \mathbb{H}) \subset \mathcal{A}(\Gamma \backslash \mathbb{H})$. The space $\mathcal{B}(\Gamma \backslash \mathbb{H})$ is dense in $\mathfrak{L}(\Gamma \backslash \mathbb{H})$, but $\mathcal{E}(\Gamma \backslash \mathbb{H})$ need not be.

3.3. Cusp forms.

Let us examine the orthogonal complement to $\mathcal{E}(\Gamma \backslash \mathbb{H})$ in $\mathcal{B}(\Gamma \backslash \mathbb{H})$. Take $f \in \mathcal{B}(\Gamma \backslash \mathbb{H})$, $E_{\mathfrak{a}}(*|\psi) \in \mathcal{E}(\Gamma \backslash \mathbb{H})$ and compute the inner product

$$\langle f, E_{\mathfrak{a}}(*|\psi) \rangle = \int_F f(z) \sum_{\gamma \in \Gamma_{\mathfrak{a}} \backslash \Gamma} \overline{\psi}(\operatorname{Im} \sigma_{\mathfrak{a}}^{-1} \gamma z) \, d\mu z \,.$$

Interchange the summation with the integration, make the substitution $z \mapsto \gamma^{-1} \sigma_{\mathfrak{a}} z$ and use the automorphy of f to get

$$\langle f, E_{\mathfrak{a}}(*|\psi) \rangle = \sum_{\gamma \in \Gamma_{\mathfrak{a}} \backslash \Gamma} \int_{\sigma_{\mathfrak{a}}^{-1} \gamma F} f(\sigma_{\mathfrak{a}} z) \, \overline{\psi}(y) \, d\mu z \,.$$

Here, as γ runs over $\Gamma_{\mathfrak{a}}\backslash\Gamma$ the sets $\sigma_{\mathfrak{a}}^{-1}\gamma F$ cover the strip $P = \{z \in \mathbb{H} : 0 < x < 1\}$ once (for an appropriate choice of representatives), giving

$$\int_P f(\sigma_{\mathfrak{a}}z)\,\overline{\psi}(y)\,d\mu z = \int_0^{+\infty} \left(\int_0^1 f(\sigma_{\mathfrak{a}}z)\,dx\right)\overline{\psi}(y)\,y^{-2}\,dy\,.$$

The innermost integral is just the zero-th term $f_{\mathfrak{a}}(y)$ in the Fourier expansion (3.2) of f at the cusp $\mathfrak{a}$. In fact, the above argument is valid for any $f \in \mathcal{A}(\Gamma\backslash\mathbb{H})$ such that $|f|$ is integrable over F. Hence we obtain

Lemma 3.3. *Let $f(z)$ be an automorphic function absolutely integrable over F. Let $E_{\mathfrak{a}}(z|\psi)$ be an incomplete Eisenstein series associated with the cusp $\mathfrak{a}$ and a test function $\psi \in C_0^\infty(\mathbb{R}^+)$. Then we have*

$$(3.14) \qquad \langle f, E_{\mathfrak{a}}(*|\psi)\rangle = \int_0^{+\infty} f_{\mathfrak{a}}(y)\,\overline{\psi}(y)\,y^{-2}\,dy,$$

where $f_{\mathfrak{a}}(y)$ is the zero-th term in the Fourier expansion of f at $\mathfrak{a}$.

Now suppose $f \in \mathcal{B}(\Gamma\backslash\mathbb{H})$ is orthogonal to the space $\mathcal{E}(\Gamma\backslash\mathbb{H})$. Then the integral (3.14) vanishes for all $\psi \in C_0^\infty(\mathbb{R}^+)$. This is equivalent with the condition

$$(3.15) \qquad f_{\mathfrak{a}}(y) \equiv 0\,, \qquad \text{for any cusp } \mathfrak{a}.$$

Denote by $\mathcal{C}(\Gamma\backslash\mathbb{H})$ the space of smooth, bounded automorphic functions whose zero-th terms at all cusps vanish. Therefore, we have just proved the orthogonal decomposition

$$(3.16) \qquad \mathfrak{L}(\Gamma\backslash\mathbb{H}) = \widetilde{\mathcal{C}}(\Gamma\backslash\mathbb{H}) \oplus \widetilde{\mathcal{E}}(\Gamma\backslash\mathbb{H}),$$

where the tilde stands for closure in the Hilbert space $\mathfrak{L}(\Gamma\backslash\mathbb{H})$ (with respect to the norm topology).

Automorphic forms in the space $\mathcal{C}(\Gamma\backslash\mathbb{H})$ are called *cusp forms*. Therefore, a cusp form f is an automorphic function which is an eigenfunction of the Laplace operator and which has no zero-th term in the Fourier expansion at any cusp, *i.e.* (3.15) holds. We denote

$$\mathcal{C}_s(\Gamma\backslash\mathbb{H}) = \mathcal{C}(\Gamma\backslash\mathbb{H}) \cap \mathcal{A}_s(\Gamma\backslash\mathbb{H}),$$

the space of cusp forms with eigenvalue $\lambda = s(1-s)$. Every $f \in \mathcal{C}_s(\Gamma\backslash\mathbb{H})$ has the expansion

$$f(\sigma_{\mathfrak{a}}z) = \sum_{n \neq 0} \hat{f}_{\mathfrak{a}}(n)\,W_s(nz)$$

at any cusp $\mathfrak{a}$, by Theorem 3.1. From the estimate (3.7) it follows that f decays exponentially at every cusp; more precisely, it satisfies

$$f(z) \ll e^{-2\pi y(z)}.$$

In particular, this shows that a cusp form is bounded on $\mathbb{H}$. Also, by Theorem 3.2 the Fourier coefficients $\hat{f}_{\mathfrak{a}}(n)$ of a cusp form are bounded. They will be the subject of intensive study in forthcoming chapters.

Clearly, $\Delta : \mathcal{C}(\Gamma \backslash \mathbb{H}) \longrightarrow \mathcal{C}(\Gamma \backslash \mathbb{H})$ and $\Delta : \mathcal{E}(\Gamma \backslash \mathbb{H}) \longrightarrow \mathcal{E}(\Gamma \backslash \mathbb{H})$. It will be shown that Δ has a pure point spectrum in $\mathcal{C}(\Gamma \backslash \mathbb{H})$, *i.e.* the space $\mathcal{C}(\Gamma \backslash \mathbb{H})$ is spanned by cusp forms. This will be accomplished by means of compact integral operators. On the other hand, in the space $\mathcal{E}(\Gamma \backslash \mathbb{H})$ the spectrum will turn out to be continuous except for a finite dimensional subspace of point spectrum. Here the analytic continuation of the Eisenstein series is the key issue. After this is established, the spectral resolution of Δ in $\mathcal{E}(\Gamma \backslash \mathbb{H})$ will evolve from (3.12) at once by contour integration. The eigenpacket of the continuous spectrum consists of the Eisenstein series $E_{\mathfrak{a}}(z, s)$ on the line $\mathrm{Re}\, s = 1/2$ (analytically continued), and the point spectrum subspace is spanned by the residues of $E_{\mathfrak{a}}(z, s)$ at poles on the segment $1/2 < s \leq 1$.

3.4. Fourier expansion of the Eisenstein series.

We begin by expanding a general Poincaré series $E_{\mathfrak{a}}(z|p)$ associated with a cusp $\mathfrak{a}$ and a test function p, see (3.1). Let $\mathfrak{b}$ be another cusp, not necessarily different from $\mathfrak{a}$. Applying the double coset decomposition (2.21) we infer that

$$E_{\mathfrak{a}}(\sigma_{\mathfrak{b}} z | p) = \sum_{\gamma \in \Gamma_{\mathfrak{a}} \backslash \Gamma} p(\sigma_{\mathfrak{a}}^{-1} \gamma \sigma_{\mathfrak{b}} z) = \sum_{\tau \in B \backslash \sigma_{\mathfrak{a}}^{-1} \Gamma \sigma_{\mathfrak{b}}} p(\tau z)$$

$$= \delta_{\mathfrak{a}\mathfrak{b}} p(z) + \sum_{c > 0} \sum_{d(\mathrm{mod}\, c)} \sum_{n \in \mathbb{Z}} p(\omega_{cd}(z + n)),$$

where the first term above exists only if $\mathfrak{a}$ and $\mathfrak{b}$ are equal. To the innermost sum we apply Poisson's formula

$$\sum_{n \in \mathbb{Z}} p(\omega_{cd}(z + n)) = \sum_{n \in \mathbb{Z}} \int_{-\infty}^{+\infty} p(\omega_{cd}(z + t)) \, e(-nt) \, dt,$$

where

$$\omega_{cd}(z + t) = a/c - c^{-2}(t + x + d/c + iy)^{-1}.$$

Changing $t \mapsto t - x - d/c$, the Fourier integral becomes

$$e(nx + n\frac{d}{c}) \int_{-\infty}^{+\infty} p(\frac{a}{c} - \frac{1}{c^2(t + iy)}) \, e(-nt) \, dt.$$

In the special case $p(z) = \psi(y)\, e(mz)$ where m is a non-negative integer we get

$$e\big(nx + n\frac{d}{c} + m\frac{a}{c}\big) \int_{-\infty}^{+\infty} \psi\big(\frac{y\,c^{-2}}{t^2 + y^2}\big)\, e\big(\frac{-m\,c^{-2}}{t + iy} - nt\big)\, dt\,.$$

Summing over the coset representatives c, d, we encounter the Kloosterman sum $\mathcal{S}_{\mathfrak{ab}}(m, n; c)$ and obtain the Fourier expansion

$$E_{\mathfrak{a}m}(\sigma_{\mathfrak{b}}z|\psi) = \delta_{\mathfrak{ab}}\, e(mz)\, \psi(y)$$

(3.17)
$$+ \sum_{n \in \mathbb{Z}} e(nx) \sum_{c > 0} \mathcal{S}_{\mathfrak{ab}}(m, n; c) \int_{-\infty}^{+\infty} \psi\big(\frac{y\,c^{-2}}{t^2 + y^2}\big)\, e\big(\frac{-m\,c^{-2}}{t + iy} - nt\big)\, dt,$$

where $\delta_{\mathfrak{ab}}$ is the diagonal symbol of Kronecker.

It remains to compute the integral in (3.17). We give results only for $m = 0$ and $\psi(y) = y^s$, which is the case of the Eisenstein series. We have

(3.18)
$$\int_{-\infty}^{+\infty} (t^2 + y^2)^{-s}\, dt = \pi^{1/2}\, \frac{\Gamma(s - 1/2)}{\Gamma(s)}\, y^{1-2s}$$

and

(3.19)
$$\int_{-\infty}^{+\infty} (t^2 + y^2)^{-s}\, e(-nt)\, dt$$
$$= 2\pi^s \Gamma(s)^{-1} |n|^{s-1/2} y^{-s+1/2} K_{s-1/2}(2\pi |n| y)$$

for $n \neq 0$ (see Appendix B). Substituting these evaluations into (3.17), we arrive at the explicit Fourier expansion of the Eisenstein series.

Theorem 3.4. *Let $\mathfrak{a}$, $\mathfrak{b}$ be cusps for Γ, and let $\operatorname{Re} s > 1$. We have*

(3.20)
$$E_{\mathfrak{a}}(\sigma_{\mathfrak{b}}z, s) = \delta_{\mathfrak{ab}}\, y^s + \varphi_{\mathfrak{ab}}(s)\, y^{1-s} + \sum_{n \neq 0} \varphi_{\mathfrak{ab}}(n, s)\, W_s(n\,z),$$

where

(3.21)
$$\varphi_{\mathfrak{ab}}(s) = \pi^{1/2}\, \frac{\Gamma(s - 1/2)}{\Gamma(s)} \sum_c c^{-2s}\, \mathcal{S}_{\mathfrak{ab}}(0, 0; c)\,,$$

(3.22)
$$\varphi_{\mathfrak{ab}}(n, s) = \pi^s \Gamma(s)^{-1} |n|^{s-1} \sum_c c^{-2s} \mathcal{S}_{\mathfrak{ab}}(0, n; c)\,,$$

and $W_s(z)$ is the Whittaker function given by (1.26).

By the trivial estimate (2.38) for Kloosterman sums and the crude bound $W_s(z) \ll \min\{y^{1-\sigma}, e^{-2\pi y}\}$ we infer the following:

Corollary 3.5. *For s on the line $\mathrm{Re}\, s = \sigma > 1$ we have*

$$(3.23) \qquad E_{\mathfrak{a}}(\sigma_{\mathfrak{b}} z, s) = \delta_{\mathfrak{a}\mathfrak{b}}\, y^s + \varphi_{\mathfrak{a}\mathfrak{b}}(s)\, y^{1-s} + O\big((1 + y^{-\sigma})\, e^{-2\pi y}\big)$$

uniformly in $z \in \mathbb{H}$, the implied constant depending on s and the group.

We have mentioned in the conclusion of the previous section that the analytic continuation of $E_{\mathfrak{a}}(z, s)$ to $\mathrm{Re}\, s \geq 1/2$ would be required for the spectral decomposition of the space $\mathcal{E}(\Gamma \backslash \mathbb{H})$. For some groups the continuation can be deduced from the Fourier expansion; we shall show it for the modular group. By (3.21) and (2.28) we get

$$(3.24) \qquad \varphi(s) = \pi^{1/2}\, \frac{\Gamma(s - 1/2)}{\Gamma(s)}\, \frac{\zeta(2s - 1)}{\zeta(2s)},$$

where $\zeta(s)$ is the Riemann zeta-function. By (3.22) and (2.27) we get

$$(3.25) \qquad \varphi(n, s) = \pi^s \Gamma(s)^{-1} \zeta(2s)^{-1} |n|^{-1/2} \sum_{ab = |n|} \left(\frac{a}{b}\right)^{s - 1/2}$$

(since there is only one cusp $\mathfrak{a} = \mathfrak{b} = \infty$ we have dropped the subscript $\mathfrak{a}\mathfrak{b}$ for simplicity).

Since the Fourier coefficients $\varphi(s)$, $\varphi(n, s)$ are meromorphic in the whole complex s-plane and the Whittaker function $W_s(z)$ is entire in the s-variable, the Fourier expansion (3.20) furnishes the meromorphic continuation of $E(z, s)$ to all $s \in \mathbb{C}$. In the half-plane $\mathrm{Re}\, s \geq 1/2$ there is only one simple pole at $s = 1$ with constant residue

$$(3.26) \qquad \mathop{\mathrm{res}}_{s=1} E(z, s) = \frac{3}{\pi}\,.$$

Note that the holomorphy of all $\varphi(n, s)$ on the line $\mathrm{Re}\, s = 1/2$ is equivalent to the non-vanishing of $\zeta(s)$ on the line $\mathrm{Re}\, s = 1$, the latter fact being equivalent to the Prime Number Theorem. To obtain some symmetry, we put

$$(3.27) \qquad \theta(s) = \pi^{-s}\, \Gamma(s)\, \zeta(2s),$$

so that

$$(3.28) \qquad \varphi(s) = \theta(1 - s)\, \theta(s)^{-1}$$

by the functional equation for the Riemann zeta-function. Now we can write the Fourier expansion (3.20) in the elegant fashion

$$(3.29) \qquad \begin{aligned} \theta(s)\, E(z, s) &= \theta(s)\, y^s + \theta(1 - s)\, y^{1-s} \\ &\quad + 4\sqrt{y} \sum_{n=1}^{\infty} \eta_{s-1/2}(n)\, K_{s-1/2}(2\pi n y)\, \cos(2\pi n x), \end{aligned}$$

where

(3.30)
$$\eta_t(n) = \sum_{ab=n} \left(\frac{a}{b}\right)^t.$$

Since the right side of (3.29) is invariant under the change $s \mapsto 1 - s$, it yields the following functional equation:

(3.31)
$$\theta(s)\,E(z,s) = \theta(1-s)\,E(z,1-s).$$

Show, using the functional equation, that $E(z, 1/2) \equiv 0$. Then evaluate (3.29) at $s = 1/2$ to get the following expansion involving the divisor function $\tau(n)$:

$$\frac{\partial}{\partial s}\,E(z,\frac{1}{2}) = \sqrt{y}\,\log y + 4\sqrt{y}\sum_{n=1}^{\infty}\tau(n)\,K_0(2\pi ny)\,\cos(2\pi nx).$$

Chapter 4

The Spectral Theorem. Discrete Part

4.1. The automorphic Laplacian.

The Laplace operator Δ acts on all smooth automorphic functions. However, for the purpose of the spectral decomposition of $\mathfrak{L}(\Gamma\backslash\mathbb{H})$ we choose the initial domain of Δ to be

$$\mathcal{D}(\Gamma\backslash\mathbb{H}) = \left\{f \in \mathcal{B}(\Gamma\backslash\mathbb{H}) : \ \Delta f \in \mathcal{B}(\Gamma\backslash\mathbb{H})\right\},$$

which is dense in $\mathfrak{L}(\Gamma\backslash\mathbb{H})$. We shall show that Δ is symmetric and non-negative. Therefore, by Theorem A.3, Δ has a unique self-adjoint extension to $\mathfrak{L}(\Gamma\backslash\mathbb{H})$.

By Stokes' theorem we have

$$\int_F \Delta f \, \overline{g} \, d\mu z = - \int_F \nabla f \, \overline{\nabla g} \, dx \, dy + \int_{\partial F} \frac{\partial f}{\partial n} \overline{g} \, d\ell,$$

where F is a bounded domain in $\mathbb{R}^2$ with a continuous and piecewise smooth boundary ∂F, f, g are smooth functions, $\nabla f = [\partial f/\partial x, \partial f/\partial y]$ is the gradient of f, $\partial f/\partial n$ is the outer normal derivative and $d\ell$ is the euclidean length element. For $F \subset \mathbb{H}$ the boundary integral can be given a hyperbolic invariant form

$$\int_{\partial F} \frac{\partial f}{\partial \mathbf{n}} \overline{g} \, d\boldsymbol{\ell},$$

where $\partial/\partial\mathbf{n} = y\partial/\partial n$ and $d\boldsymbol{\ell} = y^{-1} \, d\ell$ are G-invariant. In this form the Stokes formula remains valid for any polygon $F \subset \mathbb{H}$ of finite area provided $f, g, \Delta f, \Delta g$ are all bounded. Letting F be a fundamental polygon for a group Γ, we find that the boundary integral vanishes because the integrals along equivalent sides cancel out. Therefore we obtain

Lemma 4.1. *For* $f, g \in \mathcal{D}(\Gamma \backslash \mathbb{H})$ *we have*

$$(4.1) \qquad \langle -\Delta f, g \rangle = \int_F \nabla f \, \overline{\nabla g} \, dx \, dy,$$

whence

$$(4.2) \qquad \langle \Delta f, g \rangle = \langle f, \Delta g \rangle,$$

so Δ *is symmetric. Moreover,*

$$(4.3) \qquad \langle -\Delta f, f \rangle = \int_F |\nabla f|^2 \, dx \, dy \geq 0,$$

so $-\Delta$ *is non-negative.*

Observe that the quantity $y^2 \, \nabla f \, \overline{\nabla g} = y^2 (f_x \, \overline{g_x} + f_y \, \overline{g_y})$ is G-invariant, so the integral (4.1) does not depend on the choice of a fundamental domain.

From Lemma 4.1 it follows that the eigenvalue $\lambda = s(1 - s)$ of an eigenfunction $f \in \mathcal{D}(\Gamma \backslash \mathbb{H})$ is real and non-negative. Therefore, either $s = 1/2 + it$ with $t \in \mathbb{R}$, or $0 \leq s \leq 1$.

4.2. Invariant integral operators on $\mathcal{C}(\Gamma \backslash \mathbb{H})$.

The spectral resolution of Δ in $\mathcal{C}(\Gamma \backslash \mathbb{H})$ will be performed by means of invariant integral operators. Recall that such an operator is given by a point-pair invariant kernel $k(z, w) = k(u(z, w))$, which yields

$$(Lf)(z) = \int_{\mathbb{H}} k(z, w) \, f(w) \, d\mu w$$

for $f : \mathbb{H} \longrightarrow \mathbb{C}$. Restricting the domain of L to automorphic functions, we can write

$$(Lf)(z) = \int_F K(z, w) \, f(w) \, d\mu w$$

for $f \in \mathcal{A}(\Gamma \backslash \mathbb{H})$, where F is a fixed (once and for all) fundamental domain of Γ and the new kernel is given by the series

$$(4.4) \qquad K(z, w) = \sum_{\gamma \in \Gamma} k(z, \gamma w) \, .$$

This is called the *automorphic kernel.* Here and below we require the absolute convergence of all relevant series and integrals.

First we assume that $k(u)$ is smooth and compactly supported on $\mathbb{R}^+$. After this case is worked out, we shall replace the compactness by a weaker condition, applying a suitable approximation. Clearly

$$L : \mathcal{B}(\Gamma\backslash\mathbb{H}) \longrightarrow \mathcal{B}(\Gamma\backslash\mathbb{H}),$$

because $g(z) = (Lf)(z)$, where $f \in \mathcal{B}(\Gamma\backslash\mathbb{H})$, is bounded by the area of a circle centered at z of a fixed radius. Let us compute the zero-th term of g at a cusp $\mathfrak{a}$ (see (3.2)):

$$g_{\mathfrak{a}}(y) = \int_0^1 g(\sigma_{\mathfrak{a}} n(t)z) \, dt = \int_0^1 \left(\int_{\mathbb{H}} k(\sigma_{\mathfrak{a}} n(t)z, w) \, f(w) \, d\mu w \right) dt$$

$$= \int_{\mathbb{H}} k(z, w) \left(\int_0^1 f(\sigma_{\mathfrak{a}} n(t)w) \, dt \right) d\mu w = \int_{\mathbb{H}} k(z, w) \, f_{\mathfrak{a}}(\operatorname{Im} w) \, d\mu w,$$

where $f_{\mathfrak{a}}(y)$ is the zero-th term in the Fourier expansion of f at $\mathfrak{a}$. Hence, if $f_{\mathfrak{a}}(y)$ is identically zero, then so is $g_{\mathfrak{a}}(y)$. This proves

Proposition. *An invariant integral operator L maps the subspace $\mathcal{C}(\Gamma\backslash\mathbb{H})$ of $\mathcal{B}(\Gamma\backslash\mathbb{H})$ into itself,*

$$L : \mathcal{C}(\Gamma\backslash\mathbb{H}) \longrightarrow \mathcal{C}(\Gamma\backslash\mathbb{H}).$$

Next we examine the automorphic kernel $K(z, w)$. Unfortunately, $K(z, w)$ is not bounded on $F \times F$, no matter how small you make the support of $k(u)$. The reason is that when z, w approach the same cusp, the number of terms which count in (4.4) grows to infinity. In order to get a bounded kernel we shall subtract from $K(z, w)$ the so-called "principal parts"

$$(4.5) \qquad H_{\mathfrak{a}}(z, w) = \sum_{\gamma \in \Gamma_{\mathfrak{a}}\backslash\Gamma} \int_{-\infty}^{+\infty} k(z, \sigma_{\mathfrak{a}} n(t)\sigma_{\mathfrak{a}}^{-1}\gamma w) \, dt \, .$$

Clearly $H_{\mathfrak{a}}(z, w)$ is a well defined automorphic function in the second variable. An important fact is that the principal parts do not alter the action of L on $\mathcal{C}(\Gamma\backslash\mathbb{H})$ (see Corollary 4.4 and the remarks after it).

Lemma 4.2. *For $z, w \in \mathbb{H}$ we have uniformly*

$$(4.6) \qquad H_{\mathfrak{a}}(\sigma_{\mathfrak{a}}z, w) \ll 1 + \operatorname{Im} z \, .$$

Proof. Changing w into $\sigma_{\mathfrak{a}}w$, we need to estimate

$$H_{\mathfrak{a}}(\sigma_{\mathfrak{a}}z, \sigma_{\mathfrak{a}}w) = \sum_{\tau \in B\backslash\sigma_{\mathfrak{a}}^{-1}\Gamma\sigma_{\mathfrak{a}}} \int_{-\infty}^{+\infty} k(z, t + \tau w) \, dt \, .$$

Since $k(u)$ has compact support, the ranges of integration and summation are restricted by $|z - t - \tau w|^2 \ll \operatorname{Im} z \operatorname{Im} \tau w$. This shows that $\operatorname{Im} z \asymp \operatorname{Im} \tau w$, and the integral is bounded by $O(\operatorname{Im} z)$. By Lemma 2.10 we conclude that

$$H_{\mathfrak{a}}(\sigma_{\mathfrak{a}} z, \sigma_{\mathfrak{a}} w) \ll \left(1 + \frac{1}{\operatorname{Im} z}\right) \operatorname{Im} z = 1 + \operatorname{Im} z .$$

Lemma 4.2 shows that $H_{\mathfrak{a}}(z, w)$ is bounded in the second variable:

$$(4.7) \qquad\qquad H_{\mathfrak{a}}(z, \cdot) \in \mathcal{B}(\Gamma \backslash \mathbb{H}) .$$

Proposition 4.3. *Given* $z \in \mathbb{H}$, *the principal part* $H_{\mathfrak{a}}(z, w)$ *as a function in* w *is orthogonal to the space* $\mathcal{C}(\Gamma \backslash \mathbb{H})$, *i.e.*

$$(4.8) \qquad\qquad \langle H_{\mathfrak{a}}(z, \cdot), f \rangle = 0 \qquad \textit{if } f \in \mathcal{C}(\Gamma \backslash \mathbb{H}) .$$

Proof. Changing z into $\sigma_{\mathfrak{a}} z$, we obtain by unfolding the integral over the fundamental domain that

$$\langle H_{\mathfrak{a}}(\sigma_{\mathfrak{a}} z, \cdot), f \rangle = \int_0^{+\infty} \int_0^1 \left(\int_{-\infty}^{+\infty} k(z, n(t)w) \, dt\right) \overline{f}\left(\sigma_{\mathfrak{a}} w\right) d\mu w$$

$$= \int_0^{+\infty} \left(\int_{-\infty}^{+\infty} k(z, t + iv) \, dt\right) \left(\int_0^1 \overline{f}\left(\sigma_{\mathfrak{a}} w\right) du\right) v^{-2} \, dv,$$

where $w = u + iv$. The last integral is equal to $f_{\mathfrak{a}}(v)$, so it vanishes, proving (4.8).

We define the total "principal part" of the kernel $K(z, w)$ by adding all $H_{\mathfrak{a}}(z, w)$ over inequivalent cusps:

$$(4.9) \qquad\qquad H(z, w) = \sum_{\mathfrak{a}} H_{\mathfrak{a}}(z, w) .$$

Then we subtract $H(z, w)$ from $K(z, w)$ and call the difference

$$(4.10) \qquad\qquad \hat{K}(z, w) = K(z, w) - H(z, w)$$

the "compact part" of $K(z, w)$. This becomes a kernel on $F \times F$ of an integral operator $\hat{L}$, say, acting on functions $f : F \longrightarrow \mathbb{C}$. From Proposition 4.3 we obtain

Corollary 4.4. *For* $f \in \mathcal{C}(\Gamma \backslash \mathbb{H})$ *we have* $Lf = \hat{L}f$.

Proposition 4.5. *Let F be a fundamental polygon for Γ whose cuspidal vertices are all distinct* mod Γ. *Then the kernel $\hat{K}(z,w)$ is bounded on $F \times F$.*

Proof. As γ ranges over non-parabolic motions, the points z, γw are separated by an arbitrarily large distance for almost all γ uniformly in $z, w \in F$. To estimate this separation, consider the tesselation of $\mathbb{H}$ by copies of a fixed fundamental polygon whose parabolic vertices (cusps) are not equivalent. Therefore, since $k(u)$ is compactly supported, we have

$$K(z,w) = \sum_{\gamma \text{ parabolic}} k(z, \gamma w) + O(1).$$

Similarly, using Lemma 4.2, one shows that all terms in (4.5) give a uniformly bounded contribution except for $\gamma = 1$, so that

$$H_{\mathfrak{a}}(z,w) = \int_{-\infty}^{+\infty} k(z, \sigma_{\mathfrak{a}} n(t) \sigma_{\mathfrak{a}}^{-1} w)\, dt + O(1).$$

Combining the two estimates, we can write

$$\hat{K}(z,w) = \sum_{\mathfrak{a}} J_{\mathfrak{a}}(z,w) + O(1),$$

where $J_{\mathfrak{a}}(z,w)$ is defined by

$$J_{\mathfrak{a}}(z,w) = \sum_{\gamma \in \Gamma_{\mathfrak{a}}} k(z, \gamma w) - \int_{-\infty}^{+\infty} k(z, \sigma_{\mathfrak{a}} n(t) \sigma_{\mathfrak{a}}^{-1} w)\, dt.$$

It remains to show that $J_{\mathfrak{a}}(z,w)$ is bounded on $F \times F$. Actually we shall prove that $J_{\mathfrak{a}}(z,w)$ is bounded on $\mathbb{H} \times \mathbb{H}$. This is the crucial part of the proof. We apply the Euler-MacLaurin formula

$$\sum_{b \in \mathbb{Z}} F(b) = \int F(t)\, dt + \int \psi(t)\, dF(t),$$

where $\psi(t) = t - [t] - 1/2$, getting

$$J_{\mathfrak{a}}(\sigma_{\mathfrak{a}} z, \sigma_{\mathfrak{a}} w) = \sum_{b \in \mathbb{Z}} k(z, w+b) - \int_{-\infty}^{+\infty} k(z, w+t)\, dt$$

$$= \int_{-\infty}^{+\infty} \psi(t)\, dk(z, w+t) \ll \int_{0}^{+\infty} |k'(u)|\, du \ll 1.$$

Remarks. The basic results established in this section continue to hold true for kernels $k(u)$ which are not necessarily compactly supported but decay fast, a sufficient condition being that

$$(4.11) \qquad\qquad k(u)\,, k'(u) \ll (u+1)^{-2}\,.$$

Such a generalization can be derived from the compact case by a suitable approximation or by refining the above estimates. One should also realize that $H_{\mathfrak{a}}(z,w)$ is an incomplete Eisenstein series in the second variable. Indeed, we have

$$(4.12) \qquad\qquad \int_{-\infty}^{+\infty} k(z, n(t)z')\, dt = \sqrt{y'y}\, g(\log \frac{y'}{y}),$$

with $g(r)$ given by (1.62); hence

$$H_{\mathfrak{a}}(\sigma_{\mathfrak{a}} z, w) = \sum_{\gamma \in \Gamma_{\mathfrak{a}}\backslash\Gamma} \psi(\operatorname{Im} \sigma_{\mathfrak{a}}^{-1} \gamma w) = E_{\mathfrak{a}}(w|\psi),$$

where $\psi(v) = \sqrt{vy}\, g(\log(v/y))$. Although $\psi(v)$ might not be compactly supported, it decays quite rapidly. Therefore, Proposition 4.3 comes straight from the definition of the space $\mathcal{C}(\Gamma\backslash\mathbb{H})$.

4.3. Spectral resolution of Δ in $\mathcal{C}(\Gamma\backslash\mathbb{H})$.

To this end we shall employ a proper invariant integral operator L. By Proposition 4.5 the modified operator $\hat{L}$ is of Hilbert-Schmidt type on $\mathfrak{L}^2(F)$; in fact $\hat{L}$ has a bounded kernel. Therefore, the Hilbert-Schmidt theorem applies to $\hat{L}$ (see Appendix A.3). Any function from the range of $\hat{L}$ has the series expansion

$$(4.13) \qquad\qquad f = \sum_{j \geq 0} \langle f, u_j \rangle\, u_j(z)\,.$$

Here $\{u_j\}_{j\geq 0}$ is any maximal orthonormal system of eigenfunctions of $\hat{L}$ in the space $\mathfrak{L}^2(F)$. But the range of $\hat{L}$ is definitely not dense in $\mathfrak{L}^2(F)$, so the spectral expansion (4.13) does not hold for all $f \in \mathfrak{L}^2(F)$. It may happen that $\hat{L}$ is the trivial operator, giving nothing but the zero function to expand.

To find a good L, consider the resolvent operator

$$-(R_s f)(z) = \int_{\mathbb{H}} G_s(u(z,w))\, f(w)\, d\mu w,$$

whose kernel $G_s(u)$ is the Green function, singular at $u = 0$. In order to kill the singularity, take the difference (Hilbert's formula for iterated resolvent)

$$(4.14) \qquad L = R_s - R_a = (s(1 - s) - a(1 - a)) \, R_s \, R_a$$

for $a > s \geq 2$. This has a kernel $k(u) = G_a(u) - G_s(u)$ which satisfies the conditions (4.11) (see Lemma 1.7). Recall that $R_s = (\Delta + s(1 - s))^{-1}$ (see Theorem 1.17); hence R_s has dense range in $\mathfrak{L}(\Gamma\backslash\mathbb{H})$, and so does L by the Hilbert formula. The modified operator $\hat{L}$ is bounded on $\mathfrak{L}^2(F)$. Although $\hat{L}$ annihilates many functions, the range of $\hat{L}$ is still dense in the subspace $\mathcal{C}(\Gamma\backslash\mathbb{H}) \subset \mathfrak{L}^2(F)$. Indeed, for $f \in \mathcal{D}(\Gamma\backslash\mathbb{H})$ we create

$$g = (s(1 - s) - a(1 - a))^{-1}(\Delta + a(1 - a))(\Delta + s(1 - s))f \in \mathcal{D}(\Gamma\backslash\mathbb{H}).$$

This satisfies $Lg = f$. Moreover, if $f \in \mathcal{C}(\Gamma\backslash\mathbb{H})$, then $g \in \mathcal{C}(\Gamma\backslash\mathbb{H})$, so by Corollary 4.4 we also get $\hat{L}g = f$. Therefore, the subspace $\mathcal{C}(\Gamma\backslash\mathbb{H}) \cap \mathcal{D}(\Gamma\backslash\mathbb{H})$ is in the range of $\hat{L}$, and it is dense in $\mathcal{C}(\Gamma\backslash\mathbb{H})$. This, together with the Hilbert-Schmidt theorem and Corollary 4.4, proves the following fact.

Proposition 4.6. *Let $L : \mathcal{D}(\Gamma\backslash\mathbb{H}) \longrightarrow \mathcal{D}(\Gamma\backslash\mathbb{H})$ be the integral operator given by (4.14). Then L maps the subspace $\mathcal{C}(\Gamma\backslash\mathbb{H})$ densely into itself where it has pure point spectrum. Let $\{u_j\}$ be a complete orthonormal system of eigenfunctions of L in $\mathcal{C}(\Gamma\backslash\mathbb{H})$. Then any $f \in \mathcal{C}(\Gamma\backslash\mathbb{H}) \cap \mathcal{D}(\Gamma\backslash\mathbb{H})$ has the expansion (4.13), which converges absolutely and uniformly on compacta.*

Since L, Δ commute and they are symmetric operators, it follows from Corollary A.9 for the space $\mathcal{H} = \mathcal{C}(\Gamma\backslash\mathbb{H})$ that L has a complete orthonormal system of eigenfunctions in $\mathcal{C}(\Gamma\backslash\mathbb{H})$ which are cusp forms. Applying Proposition 4.6 for this system, we conclude the spectral resolution of Δ in $\mathcal{C}(\Gamma\backslash\mathbb{H})$.

Theorem 4.7. *The automorphic Laplace operator Δ has pure point spectrum in $\mathcal{C}(\Gamma\backslash\mathbb{H})$, i.e. $\mathcal{C}(\Gamma\backslash\mathbb{H})$ is spanned by cusp forms. The eigenspaces have finite dimension. For any complete orthonormal system of cusp forms $\{u_j\}$, every $f \in \mathcal{C}(\Gamma\backslash\mathbb{H})$ has the expansion*

$$(4.15) \qquad f(z) = \sum_j \langle f, u_j \rangle \, u_j(z),$$

converging in the norm topology. If $f \in \mathcal{C}(\Gamma\backslash\mathbb{H}) \cap \mathcal{D}(\Gamma\backslash\mathbb{H})$, then the series converges absolutely and uniformly on compacta.

Remark. The space $\mathcal{C}(\Gamma\backslash\mathbb{H})$ is perhaps trivial for generic groups, as conjectured by Phillips and Sarnak [Ph-Sa]. For the Hecke triangle groups Γ_q the recent numerical computations by Winkler [Wi] and Hejhal [He2] provide some evidence that the even cusp forms probably do not exist already for $q = 7$, though the odd ones appear in abundance.

Chapter 5

The Automorphic Green Function

5.1. Introduction.

Recall that $G_s(z, z') = G_s(u(z, z'))$ is the Green function on $\mathbb{H}$ (on a free space, so to speak, see Section 1.9). We now consider a Green function in the context of $\Gamma \backslash \mathbb{H}$, where Γ is a finite volume group. This is constructed by the method of averaging images, giving

$$(5.1) \qquad G_s(z/z') = \sum_{\gamma \in \Gamma} G_s(z, \gamma z'), \qquad \text{if } z \not\equiv z' \pmod{\Gamma}.$$

We call $G_s(z/z')$ the automorphic Green function; it is not defined for $z \equiv z' \pmod{\Gamma}$. Here we suppose that $\operatorname{Re} s = \sigma > 1$, so the series converges absolutely by virtue of $G_s(u) \ll u^{-\sigma}$ and (2.41). To simplify notation, we have not displayed the dependence of the automorphic Green function on the group, since it is fixed throughout the analysis. However, in order to avoid confusion, we write a slash between variables of the automorphic Green function, in contrast to a comma in the case of the free space Green function.

Given $z \in \mathbb{H}$, we have (see (1.47))

$$(5.2) \qquad G_s(z/z') = -\frac{m}{2\pi} \log|z - z'| + O(1), \qquad \text{as } z' \to z,$$

where m is the order of the stability group Γ_z ($m = 1$ except for the elliptic fixed points).

The automorphic Green function is an automorphic form in each variable; it has the properties: $G_s(z/z') = G_s(z'/z) = G_s(\gamma z/\gamma' z')$ for $\gamma, \gamma' \in \Gamma$,

$z \not\equiv z' \pmod{\Gamma}$, and

$$(5.3) \qquad (\Delta_z + s(1-s))G_s(z/z') = 0.$$

The resolvent operator $-R_s$ restricted to automorphic functions is given by the kernel $G_s(z/z')$:

$$(5.4) \qquad -(R_s f)(z) = \int_F G_s(z/z')\, f(z')\, d\mu z.$$

By Theorem 1.17, R_s is the inverse to $\Delta + s(1-s)$ on the space $\mathcal{B}(\Gamma \backslash \mathbb{H})$. In fact, one can show by examining the arguments in Section 1.9 that the equation

$$(5.5) \qquad (\Delta + s(1-s))R_s f = f$$

is valid in the bigger space $\mathcal{B}_\mu(\Gamma \backslash \mathbb{H})$ of smooth automorphic functions satisfying the following growth condition at any cusp:

$$(5.6) \qquad f(\sigma_{\mathfrak{a}} z) \ll y(z)^\mu .$$

The equation (5.5) holds in $\mathcal{B}_\mu(\Gamma \backslash \mathbb{H})$ if $\operatorname{Re} s > \mu + 1$. Note that $\mathcal{B}_\mu(\Gamma \backslash \mathbb{H})$ is not a subspace of $\mathfrak{L}(\Gamma \backslash \mathbb{H})$ if $\mu \geq 1/2$.

For several applications of the automorphic Green function we need to control its growth. A delicate situation occurs near the diagonal $z \equiv z'$ $\pmod{\Gamma}$; this will be seen in a double Fourier expansion.

5.2. The Fourier expansion.

Let us begin by expanding a general automorphic kernel

$$K(z, z') = \sum_{\gamma \in \Gamma} k(z, \gamma z'),$$

where $k(z, z') = k(u(z, z'))$ is a point-pair invariant. We allow a logarithmic singularity of $k(u)$ at $u = 0$ and assume the series converges absolutely whenever $z \not\equiv z' \pmod{\Gamma}$, as in the case of the Green function. Suppose $\mathfrak{a}$, $\mathfrak{b}$ are cusps for Γ (not necessarily distinct). We split the series

$$K(\sigma_{\mathfrak{a}} z, \sigma_{\mathfrak{b}} z') = \sum_{\tau \in \sigma_{\mathfrak{a}}^{-1} \Gamma \sigma_{\mathfrak{b}}} k(z, \tau z')$$

according to the double coset decomposition of the set $\sigma_{\mathfrak{a}}^{-1} \Gamma \sigma_{\mathfrak{b}}$ given in Theorem 2.7. We obtain

$$(5.7) \qquad K(\sigma_{\mathfrak{a}} z, \sigma_{\mathfrak{b}} z') = \delta_{\mathfrak{a}\mathfrak{b}}\, K_0(z, z') + \sum_{c \in \mathcal{C}_{\mathfrak{a}\mathfrak{b}}} K_c(z, z') ,$$

where K_0 exists only if $\mathfrak{a} = \mathfrak{b}$, in which case

$$K_0(z, z') = \sum_n k(z, z' + n).$$

By Poisson's summation we obtain

$$K_0(z, z') = \sum_n e(-nx + nx') P_n(y, y'),$$

say, where

$$P_n(y, y') = \int_{-\infty}^{\infty} e(\xi n) k(iy + \xi, iy') d\xi.$$

For $c \in \mathcal{C}_{\mathfrak{ab}}$ the computation of $K_c(z, z')$ is similar. We have

$$K_c(z, z') = \sum_{d(\mathrm{mod}\ c)} \sum_{m\,n} k(z + n, \omega_{cd}(z' - m))$$

$$= \sum_{d(\mathrm{mod}\ c)} \sum_{m\,n} \iint\limits_{-\infty}^{+\infty} e(\xi n + \eta m) k(z + \xi, \omega_{cd}(z' - \eta)) \, d\xi \, d\eta$$

$$= \sum_{d(\mathrm{mod}\ c)} \sum_{m\,n} e(n(-x + \frac{a}{c}) + m(x' + \frac{d}{c}))$$

$$\cdot \iint\limits_{-\infty}^{+\infty} e(\xi n + \eta m) k(iy + \xi, \frac{-c^{-2}}{iy' - \eta}) \, d\xi \, d\eta$$

$$= c^{-2} \sum_{m\,n} \mathcal{S}_{\mathfrak{ab}}(m, n; c) \, e(-nx + mx') \, P_{n, mc^{-2}}(y, c^2 y'),$$

where $\mathcal{S}_{\mathfrak{ab}}(m, n; c)$ is the Kloosterman sum (see (2.23)) and

$$P_{n,m}(y, y') = \iint\limits_{-\infty}^{+\infty} e(\xi n + \eta m) k(iy + \xi, \frac{-1}{iy' - \eta}) \, d\xi \, d\eta.$$

It remains to compute the integrals $P_n(y, y')$ and $P_{n,m}(y, y')$ explicitly in terms of $k(u)$. In full generality this seems to be a hard task; however, the special case $k(u) = G_s(u)$ is all we need. There is a natural and elegant approach through the theory of the Green function of an ordinary differential equation. But it is not short; therefore, we now state and use the results before proving them at the end of this section.

Lemma 5.1. *Let $k(u) = G_s(u)$ with $\mathrm{Re}\, s > 1$. Suppose that $y' > y$. Then we have*

(5.8) $$P_0(y, y') = (2s - 1)^{-1} y^s (y')^{1-s}$$

and

(5.9) $P_n(y, y') = (4\pi|n|)^{-1} V_s(iny) W_s(iny') , \qquad n \neq 0,$

where $W_s(z)$ and $V_s(z)$ are defined by (1.26) and (1.36), respectively.

Lemma 5.2. *Let $k(u) = G_s(u)$ with $\operatorname{Re} s > 1$. Suppose that $y'y > 1$. Then we have*

(5.10) $P_{0,0}(y, y') = \dfrac{\pi^{1/2}}{2s - 1} \dfrac{\Gamma(s - 1/2)}{\Gamma(s)} (yy')^{1-s} ,$

(5.11) $P_{0,m}(y, y') = \dfrac{\pi^s}{(2s - 1)\Gamma(s)} \left(\dfrac{y}{|m|}\right)^{1-s} W_s(imy') , \quad m \neq 0,$

(5.12) $P_{n,0}(y, y') = \dfrac{\pi^s}{(2s - 1)\Gamma(s)} \left(\dfrac{y'}{|n|}\right)^{1-s} W_s(iny) , \qquad n \neq 0,$

and

(5.13) $P_{n,m}(y, y') = \dfrac{1}{2} |mn|^{-1/2} W_s(iny) W_s(imy') \cdot \begin{cases} J_{2s-1}(4\pi\sqrt{mn}), \\ I_{2s-1}(4\pi\sqrt{|mn|}), \end{cases}$

according to whether $mn > 0$ or $mn < 0$.

The above formulas are valid in limited ranges; they are applicable for all terms of the series (5.7) in the domain

(5.14) $D_{\mathfrak{ab}} = \left\{ (z, z') \in \mathbb{H} \times \mathbb{H} : \ y' > y , \ y'y > c(\mathfrak{a}, \mathfrak{b})^{-2} \right\},$

where $c(\mathfrak{a}, \mathfrak{b})$ denotes the smallest element of $\mathcal{C}_{\mathfrak{ab}}$ (see (2.22)). If $(z, z') \in D_{\mathfrak{ab}}$, then $\sigma_{\mathfrak{a}}z \not\equiv \sigma_{\mathfrak{b}}z'$ (mod Γ), so $G_s(\sigma_{\mathfrak{a}}z/\sigma_{\mathfrak{b}}z')$ is defined. By Lemmas 5.1 and 5.2 we obtain, after changing the order of summation (the series converges absolutely), the following Fourier expansion.

Theorem 5.3. *Let $\operatorname{Re} s > 1$ and $(z, z') \in D_{\mathfrak{ab}}$. We have*

$$
\begin{aligned}
G_s(\sigma_{\mathfrak{a}}z/\sigma_{\mathfrak{b}}z') = {}& (2s - 1)^{-1}\big(\delta_{\mathfrak{ab}} y^s + \varphi_{\mathfrak{ab}}(s) y^{1-s}\big)(y')^{1-s} \\
& + \delta_{\mathfrak{ab}} \sum_{n \neq 0} (4\pi|n|)^{-1} W_s(nz') \overline{V}_{\bar{s}}(nz) \\
& + (2s - 1)^{-1} y^{1-s} \sum_{m \neq 0} \varphi_{\mathfrak{ab}}(m, s) W_s(mz') \\
& + (2s - 1)^{-1} (y')^{1-s} \sum_{n \neq 0} \varphi_{\mathfrak{ab}}(n, s) \overline{W}_{\bar{s}}(nz) \\
& + \sum_{mn \neq 0} Z_s(m, n) W_s(mz') \overline{W}_{\bar{s}}(nz),
\end{aligned}
$$

(5.15)

where $\varphi_{\mathfrak{ab}}(s)$, $\varphi_{\mathfrak{ab}}(n,s)$ *are the Fourier coefficientes of the Eisenstein series of* $E_{\mathfrak{a}}(\sigma_{\mathfrak{b}}z,s)$ *(see Theorem 3.4) and* $Z_s(m,n)$ *is the Kloosterman sums zeta-function defined by*

$$(5.16) \qquad 2\sqrt{|mn|}\, Z_s(m,n) = \sum_c c^{-1} \mathcal{S}_{\mathfrak{ab}}(m,n;c) \cdot \begin{cases} J_{2s-1}\!\left(\dfrac{4\pi}{c}\sqrt{mn}\right), \\[2mm] I_{2s-1}\!\left(\dfrac{4\pi}{c}\sqrt{|mn|}\right), \end{cases}$$

according to whether $mn > 0$ *or* $mn < 0$.

5.3. An estimate for the automorphic Green function.

We shall use the Fourier expansion (5.15) to estimate $G_s(\sigma_{\mathfrak{a}}z/\sigma_{\mathfrak{b}}z')$ in cuspidal zones. From the first and the fourth lines on the right side of (5.15) we assemble $(2s-1)^{-1}(y')^{1-s}E_{\mathfrak{b}}(\sigma_{\mathfrak{a}}z,s)$ (see Theorem 3.4). To the second line we apply the asymptotics

$$W_s(mz') = \exp\left(2\pi imx' - 2\pi|m|y'\right)(1 + O(|m|^{-1})),$$
$$V_s(nz) = \exp\left(2\pi inx + 2\pi|n|y\right)(1 + O(|n|^{-1})),$$

getting

$$\sum_{n \neq 0} (4\pi|n|)^{-1}\exp\left(2\pi in(x'-x) + 2\pi|n|(y-y')(1 + O(|n|^{-1}))\right)$$

$$= \operatorname{Re} \sum_{n=1}^{\infty} (2\pi n)^{-1}\, e(n(z-z')) + O(e^{-2\pi(y'-y)})$$

$$= -\frac{1}{\pi}\,\log|1 - e(z-z')| + O(e^{-2\pi(y'-y)}).$$

The third line is estimated by

$$y^{1-\sigma}\sum_{1}^{\infty} m^{\sigma-1}e^{-2\pi my'} \ll y^{1-\sigma}e^{-2\pi y'}.$$

For the last line we need a bound on $Z_s(m,n)$. To this end we appeal to the following crude estimates for Bessel functions:

$$I_{2s-1}(y) \ll \min\{y^{2\sigma-1}, y^{-1/2}\}\, e^y,$$
$$J_{2s-1}(y) \ll \min\{y^{2\sigma-1}, y^{-1/2}\}.$$

Now, employing the trivial bounds for Kloosterman sums, we infer that

$$Z_s(m,n) \ll \exp\left(\frac{4\pi\sqrt{|mn|}}{c(\mathfrak{a},\mathfrak{b})}\right),$$

and hence the last line on the right side of (5.15) is estimated by

$$\sum_{mn \neq 0} \exp\left(\frac{4\pi\sqrt{|mn|}}{c(\mathfrak{a},\mathfrak{b})} - 2\pi|m|y' - 2\pi|n|y\right) \ll e^{-2\pi(y'+y)}$$

provided $y'y > \delta > c(\mathfrak{a},\mathfrak{b})^{-2}$. Collecting the above estimates, we obtain

Lemma 5.4. *Let* $\operatorname{Re} s > 1$ *and* $\delta > c(\mathfrak{a},\mathfrak{b})^{-2}$. *Then for* z, z' *with* $y' > y$ *and* $y'y > \delta$ *we have*

$$
\begin{aligned}
(5.17) \qquad G_s(\sigma_\mathfrak{a}z/\sigma_\mathfrak{b}z') &= (2s-1)^{-1}\,(y')^{1-s}\,E_\mathfrak{b}(\sigma_\mathfrak{a}z,s) \\
&\quad - \frac{1}{\pi}\log|1 - e(z-z')| + O(e^{-2\pi(y'-y)}).
\end{aligned}
$$

5.4. Evaluation of some integrals.

As promised, we now give proofs of Lemmas 5.1 and 5.2. We first evaluate $P_n(y,y')$ by an appeal to the theory of the Green function (see Appendix A.5), and then we apply the result to $P_{m,n}(y,y')$ (different computations can be found in [Fa]).

The Fourier integral

$$P_n(y,y') = \int_{-\infty}^{+\infty} e(\xi n)\,k(iy+\xi, iy')\,d\xi$$

has the singular kernel $k(z,z') = G_s(u(z,z'))$ (this is a function in four variables x, y, x', y' for $z = x + iy$ and $z' = x' + iy'$), yet $P_n(y,y')$ is continuous in $\mathbb{R}^+ \times \mathbb{R}^+$ including the diagonal. Clearly $P_n(y,y')$ is symmetric. Differentiating in y twice, we get

$$P'' = \int_{-\infty}^{+\infty} e(\xi n)\,k_{yy}(iy+\xi, iy')\,d\xi,$$

and integrating by parts in ξ two times, we get

$$(2\pi in)^2 P = \int_{-\infty}^{+\infty} e(\xi n)\,k_{xx}(iy+\xi, iy')\,d\xi.$$

Summing, we obtain

$$y^2(P'' - 4\pi^2 n^2 P) = \int_{-\infty}^{+\infty} e(\xi n)\,\Delta_z k(iy+\xi, iy')\,d\xi = -s(1-s)P$$

by $(\Delta_z + s(1-s))k(z,z') = 0$ (see (1.68)). This means that $P_n(y,y')$ as a function in y satisfies the Bessel differential equation

$$(5.18) \qquad P''(y) + (s(1-s)y^{-2} - 4\pi^2 n^2)\,P(y) = 0\,.$$

Next, applying (1.66) to functions of type $f(z) = e(nx)\,g(y)$, we infer that

$$T\int_{-\infty}^{+\infty} P_n(y,y')\,g(y')\,dy' = g(y)$$

for any g smooth and bounded in $\mathbb{R}^+$, where T is the second order differential operator associated with the equation (5.18). Therefore, $P_n(y,y')$ is a Green function for T.

There are two linearly independent solutions to (5.18), namely $I(y) = V_s(iny)$ and $K(y) = W_s(iny)$ (see (1.36) and (1.26), respectively), for which the Wronskian is equal to

$$W = I'K - IK' = 4\pi\,|n|\,.$$

This follows by the asymptotic formulas (see (1.38) and (1.37))

$$I(y) \sim e^{2\pi|n|y}\,, \qquad I'(y) \sim 2\pi\,|n|\,e^{2\pi|n|y}\,,$$
$$K(y) \sim e^{-2\pi|n|y}\,, \qquad K'(y) \sim -2\pi\,|n|\,e^{-2\pi|n|y}\,,$$

as $y \to +\infty$, and by the fact that the Wronskian is constant. If $n = 0$, we have a pair $I(y) = y^s$, $K(y) = y^{1-s}$ for which the Wronskian is equal to $2s - 1$.

Finally, by the theory of the Green function for the equation (5.18) it follows that $P_n(y,y')$ is given as in Lemma 5.1.

For the proof of Lemma 5.2 we express $P_{n,m}$ in terms of P_n as the integral

$$P_{n,m}(y,y') = \int_{-\infty}^{+\infty} e\Big(\frac{-\eta n}{\eta^2 + (y')^2} - \eta m\Big)\,P_n\Big(y, \frac{y'}{\eta^2 + (y')^2}\Big)\,d\eta\,.$$

Since $y > y'(\eta^2 + (y')^2)^{-1}$ for any $\eta \in \mathbb{R}$, we can apply Lemma 5.1 in the whole range of the above integration.

First consider $n = 0$. In this case

$$P_{0,m}(y,y') = (2s-1)^{-1}y^{1-s}(y')^s \int_{-\infty}^{+\infty} e(-\eta m)\,(\eta^2 + (y')^2)^{-s}\,d\eta,$$

whence we get (5.10) and (5.11) as in the Fourier expansion for the Eisenstein series (see Theorem 3.4 and the integrals preceeding its proof).

For $n \neq 0$ we have

$$P_{n,m}(y, y') = (4\pi|n|)^{-1} W_s(iny) \, y' \int_{-\infty}^{+\infty} e(\eta m y') \, V_s\left(\frac{n/y'}{\eta + i}\right) d\eta \,.$$

Hence, we shall get (5.12) and (5.13) by virtue of the following lemma:

Lemma 5.5. *Let* $\mathrm{Re}\, s > 1/2$ *and* $a \neq 0$, b *be real numbers. Then the integral*

$$(5.19) \qquad\qquad \int_{-\infty}^{+\infty} e(\eta b) \, V_s\left(\frac{a}{\eta + i}\right) d\eta$$

is equal to

$$(5.20) \qquad 4\pi(2s - 1)^{-1}\Gamma(s)^{-1}(\pi|a|)^s \qquad\qquad \text{if } b = 0\,,$$

$$(5.21) \qquad 4\pi|a|^{1/2} K_{s-1/2}(2\pi|b|) \, J_{2s-1}(4\pi\sqrt{ab}) \qquad \text{if } ab > 0\,,$$

$$(5.22) \qquad 4\pi|a|^{1/2} K_{s-1/2}(2\pi|b|) \, I_{2s-1}(4\pi\sqrt{|ab|}) \qquad \text{if } ab < 0\,.$$

Proof. We appeal to the equation

$$(\Delta + s(1 - s))V(z) = 0,$$

where $V(z) = V_s(z)$. For $z = a(\eta + i)^{-1} = a\eta(\eta^2 + 1)^{-1} - ia(\eta^2 + 1)^{-1} = ax_\eta - iay_\eta$, say, this gives

$$a^2 y_\eta^2 (V_{xx} + V_{yy}) + s(1 - s)V = 0.$$

Hence,

$$a^2 \frac{\partial^2}{\partial\eta^2} V\left(\frac{a}{\eta + i}\right) = a^2 x_\eta^2 \, V_{xx} + 2a^2 x_\eta y_\eta \, V_{xy} + a^2 y_\eta^2 \, V_{yy}$$

$$= a^2(x_\eta^2 - y_\eta^2) \, V_{xx} + 2a^2 x_\eta y_\eta \, V_{xy} - s(1 - s)V$$

$$= 4\pi^2 a^2 x_\eta' \, V - 2\pi i a^2 y_\eta' \, V_y - s(1 - s)V,$$

because $x_\eta' = y_\eta^2 - x_\eta^2$, $y_\eta' = -2\, x_\eta y_\eta$, $V_{xx} = (2\pi i)^2 V$ and $V_{xy} = 2\pi i \, V_y$. Since

$$\frac{\partial}{\partial\eta} V\left(\frac{a}{\eta + i}\right) = 2\pi i a x_\eta' \, V + a y_\eta' \, V_y,$$

we obtain

$$a^2 \frac{\partial^2}{\partial\eta^2} V\left(\frac{a}{\eta + i}\right) + s(1 - s) V\left(\frac{a}{\eta + i}\right) + 2\pi i a \frac{\partial}{\partial\eta} V\left(\frac{a}{\eta + i}\right) = 0.$$

Let $v(a)$ denote the Fourier integral (5.19) as a function of a. Integrating by parts and using the above relation, we find that $v(a)$ satisfies the second order differential equation

$$(5.23) \qquad a^2 v''(a) + (s(1-s) + 4\pi^2 ab)\, v(a) = 0.$$

If $b = 0$, the solutions to (5.23) are $v(a) = \alpha\, |a|^s + \beta\, |a|^{1-s}$, where α, β are constants. We shall determine these constants from the asymptotic formula

$$V(z) \sim 2\pi^{s+1/2}\, \Gamma(s+1/2)^{-1} |y|^s, \qquad \text{as } y \to 0,$$

which yields

$$v(a) \sim 2\pi^{s+1/2}\, \Gamma(s+1/2)^{-1} |a|^s \int_{-\infty}^{+\infty} (1+\eta^2)^{-s}\, d\eta,$$

as $a \to 0$. Hence, by (3.18) and the duplication formula for the gamma function (see (B.6) in Appendix B), we get

$$v(a) \sim \frac{4\pi(\pi|a|)^s}{(2s-1)\Gamma(s)}.$$

This shows first that $\beta = 0$, and then that $v(a)$ is exactly equal to (5.20).

If $b \neq 0$, the solutions to (5.23) are given by Bessel's functions

$$v(a) = \alpha |a|^{1/2} J_{2s-1}(4\pi\sqrt{ab}) + \beta |a|^{1/2} Y_{2s-1}(4\pi\sqrt{ab}),$$

if $ab > 0$, and

$$v(a) = \alpha |a|^{1/2} I_{2s-1}(4\pi\sqrt{|ab|}) + \beta |a|^{1/2} K_{2s-1}(4\pi\sqrt{|ab|}),$$

if $ab < 0$. From the power series expansion for Bessel's functions it follows that

$$J_{2s-1}(4\pi\sqrt{ab}) \sim \Gamma(2s)^{-1}(2\pi\sqrt{ab})^{2s-1},$$
$$I_{2s-1}(4\pi\sqrt{|ab|}) \sim \Gamma(2s)^{-1}(2\pi\sqrt{|ab|})^{2s-1},$$

and similar asymptotics hold true for Y_{2s-1} and K_{2s-1}, but with s replaced by $1-s$. On the other hand, we infer by (3.19) that

$$v(a) \sim 2\pi^{s+1/2}\, \Gamma(s+1/2)^{-1} |a|^s \int_{-\infty}^{+\infty} e(\eta b)\, (1+\eta^2)^{-s}\, d\eta$$
$$= 2\, (2\pi)^{2s}\, \Gamma(2s)^{-1} |a|^s |b|^{s-1/2} K_{s-1/2}(2\pi|b|).$$

From this, one determines the constants $\beta = 0$ and $\alpha = 4\pi\, K_{s-1/2}(4\pi|b|)$, so $v(a)$ is given exactly by (5.21) or (5.22) according to the sign of ab. This completes the proof of Lemma 5.5 and that of Lemma 5.2.

Chapter 6

Analytic Continuation of the Eisenstein Series

The analytic continuation of the Eisenstein series $E_{\mathfrak{a}}(z, s)$ is fundamental for the spectral resolution of Δ in the space $\mathcal{E}(\Gamma\backslash\mathbb{H})$ of incomplete Eisenstein series. There are many ways to perform the analytic continuation (Selberg, Faddeev, Colin de Verdière, Langlands, Bernstein, ...). We shall present one of Selberg's methods, which uses the Fredholm theory of integral equations (see [Se2]).

6.1. The Fredholm equation for the Eisenstein series.

To get started, we fix a number $a > 2$ and consider the resolvent R_a. We have

$$(6.1) \qquad -(\Delta + a(1-a))^{-1}f(z) = \int_F G_a(z/z')\, f(z')\, d\mu z',$$

for any $f \in \mathcal{B}_{a-1}(\Gamma\backslash\mathbb{H})$, where $G_a(z/z')$ is the automorphic Green function. Let $E_{\mathfrak{a}}(z, s)$ be the Eisenstein series for the cusp $\mathfrak{a}$. From the Fourier expansion

$$E_{\mathfrak{a}}(\sigma_{\mathfrak{b}}z, s) = \delta_{\mathfrak{a}\mathfrak{b}}\, y^s + \varphi_{\mathfrak{a}\mathfrak{b}}(s)\, y^{1-s} + E_{\mathfrak{a}}^*(\sigma_{\mathfrak{b}}z, s)$$

it is apparent that $E_{\mathfrak{a}}(z, s)$ belongs to $\mathcal{B}_{\sigma}(\Gamma\backslash\mathbb{H})$ with $\sigma = \operatorname{Re} s$. Suppose that $1 < \sigma \le a - 1$, so (6.1) applies to

$$f(z) = (\Delta + a(1-a))E_{\mathfrak{a}}(z, s) = \big(a(1-a) - s(1-s)\big)E_{\mathfrak{a}}(z, s),$$

giving

$$(6.2) \qquad -E_{\mathfrak{a}}(z, s) = \big(a(1-a) - s(1-s)\big) \int_F G_a(z/z')\, E_{\mathfrak{a}}(z', s)\, d\mu z'.$$

This is a homogeneous Fredholm equation for $E_{\mathfrak{a}}(z, s)$, but the classical Fredholm theory cannot be employed for several reasons.

The first obstacle is that the kernel $G_a(z/z')$ is singular on the diagonal $z = z'$. This is a minor problem. The singularities will disappear if we take the difference

$$G_{ab}(z/z') = G_a(z/z') - G_b(z/z')$$

for fixed $a > b > 2$. From (6.2) we obtain the homogeneous equation

$$(6.3) \qquad -\nu_{ab}(s)\, E_{\mathfrak{a}}(z, s) = \int_F G_{ab}(z/z')\, E_{\mathfrak{a}}(z', s)\, d\mu z',$$

where

$$\nu_{ab}(s) = \big(a(1 - a) - s(1 - s)\big)^{-1} - \big(b(1 - b) - s(1 - s)\big)^{-1}.$$

Later we shall put $\lambda_{ab}(s) = -\nu_{ab}(s)^{-1}$ on the other side of (6.3). Note that $\lambda_{ab}(s)$ is a polynomial in s of degree four,

$$(6.4) \qquad \lambda_{ab}(s) = \frac{(a - s)(a + s - 1)(b - s)(b + s - 1)}{(b - a)(a + b - 1)}.$$

The new kernel $G_{ab}(z/z')$ is continuous in $(z, z') \in \mathbb{H} \times \mathbb{H}$, since the leading term (the singular term) in the asymptotic (5.2) for $G_s(z/z')$ does not depend on s, so it cancels in the difference $G_{ab}(z/z')$.

The second obstacle is that $G_{ab}(z/z')$ is not bounded. We handle this problem in the z' variable by subtracting suitable contributions when z' is in cuspidal zones. To this end we fix a fundamental polygon F having inequivalent cuspidal vertices, and partition it into

$$F = F(Y) \cup \bigcup_{\mathfrak{b}} F_{\mathfrak{b}}(Y),$$

where Y is a large parameter, $F_{\mathfrak{b}}(Y)$ are the cuspidal zones of height Y and $F(Y)$ is the central part (see (2.1)-(2.5)). We define the truncated kernel on $\mathbb{H} \times F$ (not on $\mathbb{H} \times \mathbb{H}$) by setting $G_{ab}^Y(z/z') = G_{ab}(z/z')$ if $z' \in F(Y)$, and

$$G_{ab}^Y(z/z') = G_{ab}(z/z') - (2a - 1)^{-1}(\operatorname{Im} \sigma_{\mathfrak{b}}^{-1} z')^{1-a} E_{\mathfrak{b}}(z, a)$$
$$+ (2b - 1)^{-1}(\operatorname{Im} \sigma_{\mathfrak{b}}^{-1} z')^{1-b} E_{\mathfrak{b}}(z, b)$$

if $z' \in F_{\mathfrak{b}}(Y)$. Note that $G_{ab}^Y(z/z')$ is automorphic in z but not in z' (the second variable is confined to the fixed fundamental domain; its range could be extended over all $\mathbb{H}$ by Γ-periodicity, but there is no reason to do so). In the z' variable in F the truncated kernel $G_{ab}^Y(z/z')$ is continuous except for jumps on the horocycles $L_{\mathfrak{b}}(Y)$. As z' approaches a cusp, $G_{ab}^Y(z/z')$ decays exponentially by Lemma 5.4, but in the z variable the kernel $G_{ab}^Y(z/z')$ is not bounded; it has polynomial growth at cusps which is inherited from the Eisenstein series $E_{\mathfrak{b}}(z, a)$ and $E_{\mathfrak{b}}(z, b)$. More precisely, from the approximation (5.17) we infer the following bound:

$$(6.5) \qquad G_{ab}^Y(\sigma_{\mathfrak{a}} z/\sigma_{\mathfrak{b}} z') \ll y^a e^{-2\pi \max\{y'-y, 0\}}, \qquad \text{if } y, y' > Y.$$

Replacing $G_{ab}(z/z')$ in (6.3) by $G_{ab}^Y(z/z')$, we must bring back the integrals of subtracted quantities over cuspidal zones. They yield

$$\int_{F_{\mathfrak{b}}(Y)} (\operatorname{Im} \sigma_{\mathfrak{b}}^{-1} z')^{1-a} E_{\mathfrak{a}}(z', s) \, d\mu z'$$

$$= \int_0^1 \int_Y^{+\infty} y^{-1-a} (\delta_{\mathfrak{a}\mathfrak{b}} \, y^s + \varphi_{\mathfrak{a}\mathfrak{b}}(s) \, y^{1-s} + \cdots) \, dx \, dy$$

$$= \delta_{\mathfrak{a}\mathfrak{b}} \frac{Y^{s-a}}{a-s} + \varphi_{\mathfrak{a}\mathfrak{b}}(s) \frac{Y^{1-a-s}}{a+s-1},$$

for every $\mathfrak{b}$, and similar terms must be added with b in place of a. In this way we obtain the inhomogeneous Fredholm equation

$$
(6.6) \quad
\begin{aligned}
-\nu_{ab}(s) \, E_{\mathfrak{a}}(z, s) = {}& \int_F G_{ab}^Y(z/z') \, E_{\mathfrak{a}}(z', s) \, d\mu z' \\
& + \frac{Y^{s-a}}{(2a-1)(a-s)} E_{\mathfrak{a}}(z, a) \\
& - \frac{Y^{s-b}}{(2b-1)(b-s)} E_{\mathfrak{a}}(z, b) \\
& + \frac{Y^{1-a-s}}{(2a-1)(a+s-1)} \sum_{\mathfrak{b}} \varphi_{\mathfrak{a}\mathfrak{b}}(s) \, E_{\mathfrak{b}}(z, a) \\
& - \frac{Y^{1-b-s}}{(2b-1)(b+s-1)} \sum_{\mathfrak{b}} \varphi_{\mathfrak{a}\mathfrak{b}}(s) \, E_{\mathfrak{b}}(z, b).
\end{aligned}
$$

A new obstacle has emerged from the terms involving the scattering matrix elements $\varphi_{\mathfrak{a}\mathfrak{b}}(s)$ whose analytic continuation to $\operatorname{Re} s \le 1$ has not yet been established. We shall kill these terms by making a suitable linear combination of (6.6) for three values $Y, 2Y, 4Y$ (one could also accomplish the same goal by integrating in Y against a suitable test function such that its Mellin transform vanishes at the points $-a$ and $-b$). We find the following equation:

$$(6.7) \qquad h(z) = f(z) + \lambda \int_F H(z, z') \, h(z') \, d\mu z',$$

where $\lambda = \lambda_{ab}(s)$ is given by (6.4),

$$h(z) = (2^{2s-1} - 1)^{-1}(2^{s-1+a} - 1)\,(2^{s-1+b} - 1)\,\nu_{ab}(s)\,E_{\mathfrak{a}}(z, s)\,,$$

$$f(z) = \frac{2^{2s-1-a+b}}{(2b-1)(b-s)}\,Y^{s-b}E_{\mathfrak{a}}(z, b) - \frac{2^{2s-1-a+b}}{(2a-1)(a-s)}\,Y^{s-a}E_{\mathfrak{a}}(z, a)\,,$$

$$H(z, z') = (2^{s-1+a} - 1)^{-1}(2^{s-1+b} - 1)^{-1}$$
$$\cdot\left(G_{ab}^Y - 2^{s-1}(2^a + 2^b)G_{ab}^{2Y} + 2^{2s-2+a+b}G_{ab}^{4Y}\right)(z/z')\,.$$

For notational simplicity we did not display the dependence of $h(z)$, $f(z)$, $H(z, z')$ on the complex parameter s nor on the fixed numbers a, b.

6.2. The analytic continuation of $E_{\mathfrak{a}}(z, s)$.

We are almost ready to apply the Fredholm theory to the equation (6.7). There are still minor problems with the growth of $f(z)$ and $H(z, z')$ in the z variable. These functions are not bounded (as required in our version of the Fredholm theory), but they have polynomial growth at the cusps. More precisely, given $a > b > c + 1$, we have, uniformly, for s in the strip $1 - c \le \operatorname{Re} s \le c$,

$$f(\sigma_{\mathfrak{a}}z) \ll y^a$$

and

$$H(\sigma_{\mathfrak{a}}z, \sigma_{\mathfrak{b}}z') \ll y^a e^{-2\pi \max\{y'-y, 0\}}\,,$$

if $y, y' \ge 4Y$ by (6.5). To handle the problem we multiply (6.7) throughout by $\eta(z) = e^{-\eta y(z)}$, where η is a small positive constant, $0 < \eta < 2\pi$. Then we borrow from the exponential decay in the z' variable to kill the polynomial growth in the z variable. Accordingly, we rewrite (6.7) as follows:

$$(6.8) \quad \eta(z)\,h(z) = \eta(z)\,f(z) + \lambda \int_F \eta(z)\,\eta(z')^{-1}H(z, z')\,\eta(z')\,h(z')\,d\mu z'\,.$$

Here $\eta(z)\,f(z)$ is bounded in F, and $\eta(z)\,\eta(z')^{-1}H(z, z')$ is bounded in $F \times F$. By the Fredholm theory the kernel $\eta(z)\,\eta(z')^{-1}H(z, z')$ has a resolvent of type

$$(6.9) \quad\quad\quad R_\lambda(z, z') = \mathcal{D}(\lambda)^{-1}\,\mathcal{D}_\lambda(z, z')\,,$$

where $\mathcal{D}(\lambda) \not\equiv 0$ and $\mathcal{D}_\lambda(z, z')$ are holomorphic in λ of order ≤ 2; therefore of order ≤ 8 in s in the strip $1 - c \le \operatorname{Re} s \le c$.

For any λ with $\mathcal{D}(\lambda) \ne 0$ we have a unique solution to (6.8), given by

$$\eta(z)\,h(z) = \eta(z)\,f(z) + \lambda \int_F R_\lambda(z, z')\,\eta(z')\,f(z')\,d\mu z'\,,$$

whence

$$(6.10) \qquad h(z) = f(z) + \frac{\lambda}{\mathcal{D}(\lambda)} \int_F \eta(z)^{-1} \eta(z') \, \mathcal{D}_\lambda(z,z') \, f(z') \, d\mu z' \, .$$

The function $\mathcal{D}_\lambda(z,z')$ has a power series expansion in λ whose coefficients are bounded in $F \times F$. Therefore, (6.10) yields the analytic continuation of $E_{\mathfrak{a}}(z,s)$. Putting

$$A_{\mathfrak{a}}(s) = (2^{s+a-1} - 1)(2^{s+b-1} - 1) \, \mathcal{D}(\lambda),$$

where $\lambda = \lambda_{ab}(s)$ is given by (6.4), and

$$A_{\mathfrak{a}}(z,s) = (2^{2s-1} - 1) \, \lambda \, \mathcal{D}(\lambda) \, h(z),$$

where $h(z)$ is given by (6.10), we obtain

Proposition 6.1. *Given $c > 1$, denote $\mathcal{S} = \{s \in \mathbb{C} : 1 - c \leq \operatorname{Re} s \leq c\}$. There are functions $A_{\mathfrak{a}}(s) \not\equiv 0$ on $\mathcal{S}$ and $A_{\mathfrak{a}}(z,s)$ on $\mathbb{H} \times \mathcal{S}$ with the following properties:*

$$(6.11) \qquad A_{\mathfrak{a}}(s) \text{ is holomorphic in } s \text{ of order} \leq 8 \,,$$

$$(6.12) \qquad A_{\mathfrak{a}}(z,s) \text{ is holomorphic in } s \text{ of order} \leq 8 \,,$$

$$(6.13) \qquad A_{\mathfrak{a}}(z,s) \text{ is real-analytic in } (z,s) \,,$$

$$(6.14) \qquad A_{\mathfrak{a}}(z,s) \in \mathcal{A}_s(\Gamma \backslash \mathbb{H}) \,,$$

$$(6.15) \qquad A_{\mathfrak{a}}(z,s) = A_{\mathfrak{a}}(s) \, E_{\mathfrak{a}}(z,s) \text{ if } 1 < \operatorname{Re} s \leq c \,,$$

$$(6.16) \qquad A_{\mathfrak{a}}(z,s) \ll e^{\varepsilon \, y(z)} \,,$$

the implied constant depending on ε, s and Γ.

From (6.15) we draw the analytic continuation of $E_{\mathfrak{a}}(z,s)$ to the strip $\mathcal{S}$, and since c is arbitrary we get the meromorphic continuation to the whole s-plane. Furthermore, by (6.16)(where $\varepsilon = \eta$ from $\eta(z) = e^{-\eta y(z)}$) we retain certain control of growth, namely

Corollary 6.2. *Suppose s is not a zero of $A_{\mathfrak{a}}(s)$. For any $\varepsilon > 0$ we have*

$$(6.17) \qquad E_{\mathfrak{a}}(z,s) \ll e^{\varepsilon \, y(z)} \,,$$

the implied constant depending on ε, s and Γ.

This is not a very strong bound; nevertheless it helps us to proceed further. From (6.17) we infer the validity of the Fourier expansion

$$(6.18) \qquad E_{\mathfrak{a}}(\sigma_{\mathfrak{b}} z, s) = \delta_{\mathfrak{a}\mathfrak{b}} \, y^s + \varphi_{\mathfrak{a}\mathfrak{b}}(s) \, y^{1-s} + \sum_{n \neq 0} \varphi_{\mathfrak{a}\mathfrak{b}}(n,s) \, W_s(nz) \,,$$

for all s with $A_{\mathfrak{a}}(s) \neq 0$. We also obtain the meromorphic continuation of the coefficientes $\varphi_{\mathfrak{ab}}(s)$, $\varphi_{\mathfrak{ab}}(n, s)$. After this is known, the estimate (6.17) improves itself via the Fourier expansion (6.18). Manipulating skillfully with the exponentical decay of the Whittaker function, one shows that

$$(6.19) \qquad \varphi_{\mathfrak{ab}}(n, s) \ll |n|^{\sigma} + |n|^{1-\sigma}$$

and

$$(6.20) \qquad E_{\mathfrak{a}}(\sigma_{\mathfrak{b}}z, s) = \delta_{\mathfrak{ab}}\, y^{s} + \varphi_{\mathfrak{ab}}(s)\, y^{1-s} + O(e^{-2\pi y})$$

as $y \to +\infty$, for any s with $A_{\mathfrak{a}}(s) \neq 0$. However, the implied constants in (6.19) and (6.20) may depend badly on s.

We conclude with the following obvious, yet basic, observation.

Theorem 6.3. *The meromorphically continued Eisenstein series are orthogonal to cusp forms.*

Proof. The inner product of a cusp form against an Eisenstein series exists because the former has exponential decay at cusps and the latter has at most polynomial growth. For $\operatorname{Re} s > 1$ the orthogonality follows by the unfolding method, and for the regular points with $\operatorname{Re} s < 1$ it follows by analytic continuation (the inner product converges absolutely and uniformly on compacta in s).

6.3. The functional equations.

It may surprise anyone that the functional equations for the scattering matrix and the Eisenstein series come as consequences of the apparently remote facts that Δ is a symmetric and non-negative operator in $\mathcal{L}(\Gamma\backslash\mathbb{H})$. We shall appeal to these facts to establish the following:

Lemma 6.4. *Suppose $f \in \mathcal{A}_s(\Gamma\backslash\mathbb{H})$ satisfies the growth condition*

$$(6.21) \qquad f(z) \ll e^{\varepsilon\, y(z)},$$

with $0 < \varepsilon < 2\pi$. If $\operatorname{Re} s > 1$, then $f(z)$ is a linear combination of the Eisenstein series $E_{\mathfrak{a}}(z, s)$.

Proof. Since $f \in \mathcal{A}_s(\Gamma\backslash\mathbb{H})$, it has the Fourier expansion

$$f(\sigma_{\mathfrak{a}}z) = \alpha_{\mathfrak{a}}\, y^{s} + \beta_{\mathfrak{a}}\, y^{1-s} + O(1),$$

where the error term is shown to be bounded using the growth condition (6.21). Subtracting the Eisenstein series, we kill the leading terms $\alpha_\mathfrak{a}\, y^s$ and get

$$g(z) = f(z) - \sum_\mathfrak{a} \alpha_\mathfrak{a}\, E_\mathfrak{a}(z,s) \ll 1$$

in $\mathbb{H}$. Hence $g \in \mathcal{A}_s(\Gamma\backslash\mathbb{H}) \cap \mathfrak{L}(\Gamma\backslash\mathbb{H})$, which implies $g = 0$ because Δ has only non-negative eigenvalues in $\mathfrak{L}(\Gamma\backslash\mathbb{H})$. Therefore,

$$f(z) = \sum_\mathfrak{a} \alpha_\mathfrak{a}\, E_\mathfrak{a}(z,s).$$

Let $\mathcal{E}(z,s)$ denote the column vector of the Eisenstein series $E_\mathfrak{a}(z,s)$, where $\mathfrak{a}$ ranges over all inequivalent cusps. Recall the Fourier expansions (6.18). The first coefficients of the zero-th term form the identity matrix

$$I = \big(\delta_{\mathfrak{a}\mathfrak{b}}\big),$$

and the second coefficients form the so-called scattering matrix

$$\Phi(s) = \big(\varphi_{\mathfrak{a}\mathfrak{b}}(s)\big).$$

Theorem 6.5. *The column-vector Eisenstein series satisfies the functional equation*

$$(6.22) \qquad \mathcal{E}(z,s) = \Phi(s)\,\mathcal{E}(z,1-s)\,.$$

Proof. Suppose $\operatorname{Re} s > 1$ and $A_\mathfrak{a}(1-s) \neq 0$, so the Eisenstein series $E_\mathfrak{a}(z,1-s)$ is defined by meromorphic continuation:

$$E_\mathfrak{a}(z,1-s) \in \mathcal{A}_{1-s}(\Gamma\backslash\mathbb{H}) = \mathcal{A}_s(\Gamma\backslash\mathbb{H})\,.$$

Moreover, $E_\mathfrak{a}(z,1-s)$ satisfies the growth condition (6.21) by virtue of Corollary 6.2; therefore, by Lemma 6.4 it follows that

$$E_\mathfrak{a}(z,1-s) = \sum_\mathfrak{b} \varphi_{\mathfrak{a}\mathfrak{b}}(1-s)\, E_\mathfrak{b}(z,s)\,.$$

This relation extends to all $s \in \mathbb{C}$ by analytic continuation, so on changing s into $1-s$ we get (6.22).

From (6.22) by the symmetry of $\Phi(s)$ one gets another functional equation:

$$(6.22') \qquad {}^t\mathcal{E}(z,s)\,\mathcal{E}(w,1-s) = {}^t\mathcal{E}(z,1-s)\,\mathcal{E}(w,s)\,.$$

Theorem 6.6. *The scattering matrix satisfies the functional equation*

(6.23) $$\Phi(s)\,\Phi(1-s) = I\,.$$

For s with $\operatorname{Re} s = 1/2$ the scattering matrix is unitary,

(6.24) $$\Phi(s)\,{}^{t}\overline{\Phi}(s) = I\,.$$

For s real the scattering matris is hermitian.

Proof. The functional equation (6.23) follows by (6.22). Next, we see the symmetry

(6.25) $$\varphi_{\mathfrak{a}\mathfrak{b}}(s) = \varphi_{\mathfrak{b}\mathfrak{a}}(s)$$

from the Dirichlet series representation (3.21) if $\operatorname{Re} s > 1$, and it extends to all s by analytic continuation. In matrix notation (6.25) takes the form

(6.26) $$\Phi(s) = {}^{t}\Phi(s)\,.$$

We also read from the Dirichlet series representation that

(6.27) $$\overline{\Phi}(s) = \Phi(\bar{s})\,,$$

for all s, by analytic continuation. Since $\bar{s} = 1 - s$ on the line $\operatorname{Re} s = 1/2$, it follows by combining (6.26) with (6.27) that $\Phi(1-s) = \overline{{}^{t}\Phi(s)}$. This and the functional equation (6.23) yield (6.24). Finally, it follows from (6.26) and (6.27) that $\Phi(s)$ is hermitian for s real.

Denote by $\Phi_{\mathfrak{a}}(s) = [\ldots, \varphi_{\mathfrak{a}\mathfrak{b}}(s), \ldots]$ the $\mathfrak{a}$-th row vector of the scattering matrix $\Phi(s)$, and denote its ℓ_2-norm by

$$\|\Phi_{\mathfrak{a}}(s)\|^2 = \sum_{\mathfrak{b}} |\varphi_{\mathfrak{a}\mathfrak{b}}(s)|^2\,.$$

Since $\Phi(s)$ is unitary on the critical line (see (6.24)), it follows that $\Phi(s)$ is holomorphic on this line and

(6.28) $$\|\Phi_{\mathfrak{a}}(s)\|^2 = 1\,, \qquad \text{if } \operatorname{Re} s = 1/2\,.$$

6.4. Poles and residues of the Eisenstein series.

We shall infer some information about poles of $\Phi(s)$ and $\mathcal{E}(z,s)$ in $\operatorname{Re} s > 1/2$ from a certain formula of Maass and Selberg for the inner product of truncated Eisenstein series (the whole series $E_\mathfrak{a}(z,s)$ is not in $\mathfrak{L}(\Gamma \backslash \mathbb{H})$ because of polynomial growth at cusps). We set

$$(6.29) \qquad E_\mathfrak{a}^Y(z,s) = E_\mathfrak{a}(z,s) - \delta_{\mathfrak{ab}}\,(\mathrm{Im}\,\sigma_\mathfrak{b}^{-1}z)^s - \varphi_{\mathfrak{ab}}(s)\,(\mathrm{Im}\,\sigma_\mathfrak{b}^{-1}z)^{1-s}$$

if z is in the zone $F_\mathfrak{b}(Y)$, and we subtract nothing if z is in the central part $F(Y)$. The truncated Eisenstein series satisfies the bound

$$E_\mathfrak{a}^Y(z,s) \ll e^{-2\pi\,y(z)},$$

for $z \in F$, the implied constant depending on s and Y (see (6.20)).

Proposition 6.8 (Maass-Selberg). *If s_1, s_2 are regular points of the Eisenstein series $E_\mathfrak{a}(z,s)$ and $E_\mathfrak{b}(z,s)$, respectively, and $s_1 \neq s_2$, $s_1 + s_2 \neq 1$, then*

$$
\begin{aligned}
\langle E_\mathfrak{a}^Y(\cdot,s_1), \overline{E_\mathfrak{b}^Y(\cdot,s_2)} \rangle &= (s_1 - s_2)^{-1}\varphi_{\mathfrak{ab}}(s_2)\,Y^{s_1-s_2} \\
&\quad + (s_2 - s_1)^{-1}\varphi_{\mathfrak{ab}}(s_1)\,Y^{s_2-s_1} \\
&\quad + (s_1 + s_2 - 1)^{-1}\delta_{\mathfrak{ab}}\,Y^{s_1+s_2-1} \\
&\quad - (s_1 + s_2 - 1)^{-1}\Phi_\mathfrak{a}(s_1)\,\overline{\Phi_\mathfrak{b}(\overline{s_2})}\,Y^{1-s_1-s_2},
\end{aligned}
$$

(6.30)

where in the last term we have the scalar product of two row vectors of the scattering matrix.

This relation is derived by application of Green's formula to the central part $F(Y)$ of a fundamental polygon. The resulting boundary terms on equivalent side segments cancel out, and the remaining integrals along horocycles of height Y for each cusp are computed using the Fourier expansions (6.18). Only the zero-th terms survive the integration, and they make up the right-hand side of (6.30). A similar relation holds true for general Maass forms (see the end of this section for more details).

We shall need (6.30) for $\mathfrak{a} = \mathfrak{b}$ and $s_1 = \sigma + iv$, $s_2 = \sigma - iv$ with $\sigma > 1/2$ and $v \neq 0$. In this case we obtain

$$
\begin{aligned}
\|E_\mathfrak{a}^Y(\cdot,\sigma+iv)\|^2 &+ (2\sigma - 1)^{-1}Y^{1-2\sigma}\sum_\mathfrak{b}|\varphi_{\mathfrak{ab}}(\sigma+iv)|^2 \\
&= (2\sigma - 1)^{-1}Y^{2\sigma-1} - v^{-1}\mathrm{Im}\,\varphi_{\mathfrak{aa}}(\sigma+iv)\,Y^{-2iv},
\end{aligned}
$$

(6.31)

provided $s = \sigma + iv$ is a regular point of $\Phi_\mathfrak{a}(s)$. Hence, by examining the growth of individual terms and using the positivity of the left side, one immediately derives

Theorem 6.9. *The functions $\varphi_{\mathfrak{a}\mathfrak{b}}(s)$ are holomorphic in* $\operatorname{Re} s \geq 1/2$ *except for a finite number of simple poles in the segment* $(1/2, 1]$. *If* $s = s_j$ *is a pole of* $\varphi_{\mathfrak{a}\mathfrak{b}}(s)$, *then it is also a pole of* $\varphi_{\mathfrak{a}\mathfrak{a}}(s)$. *The residue of* $\varphi_{\mathfrak{a}\mathfrak{a}}(s)$ *at* $s = s_j > 1/2$ *is real and positive.*

Now we are ready to examine poles and residues of the Eisenstein series $E_{\mathfrak{a}}(z, s)$ in $\operatorname{Re} s > 1/2$. Suppose s_j is a pole of order $m \geq 1$. Then the function

$$u(z) = \lim_{s \to s_j} (s - s_j)^m E_{\mathfrak{a}}(z, s)$$

does not vanish identically, and it belongs to $\mathcal{A}_{s_j}(\Gamma \backslash \mathbb{H})$. Moreover, it has the Fourier expansion at any cusp of type

$$u(\sigma_{\mathfrak{b}} z) = \rho_{\mathfrak{b}} \, y^{1 - s_j} + \sum_{n \neq 0} \rho_{\mathfrak{b}}(n) \, W_{s_j}(nz)$$

with

$$\rho_{\mathfrak{b}} = \lim_{s \to s_j} (s - s_j)^m \varphi_{\mathfrak{a}\mathfrak{b}}(s) \, .$$

Note that the first part of the zero-th term in the Fourier expansion of $E_{\mathfrak{a}}(\sigma_{\mathfrak{a}} z, s)$ is killed in the limit. If $\operatorname{Re} s_j > 1/2$, then $u(z)$ is square-integrable; thus its eigenvalue must be real and non-negative, so s_j must lie in the segment $(1/2, 1]$. Moreover, if s_j were not a pole of $\Phi_{\mathfrak{a}}(s)$ or if s_j had order $m > 1$, then $\rho_{\mathfrak{b}} = 0$ for any $\mathfrak{b}$, showing that $u(z)$ is a cusp form. This, however, is impossible, because the Eisenstein series $E_{\mathfrak{a}}(z, s)$ with $s \neq s_j$ is orthogonal to cusp forms; hence the limit $u(z)$ would be orthogonal to itself. We conclude the above analysis by

Theorem 6.10. *The poles of $E_{\mathfrak{a}}(z, s)$ in $\operatorname{Re} s > 1/2$ are among the poles of $\varphi_{\mathfrak{a}\mathfrak{a}}(s)$, and they are simple. The residues are Maass forms; they are square-integrable on F and orthogonal to cusp forms.*

Next we determine what happens on the line $\operatorname{Re} s = 1/2$. We let $\sigma \to 1/2$ in (6.31), showing that

(6.32) $$\| E_{\mathfrak{a}}^Y (\cdot, \sigma + iv) \| \ll 1 \, ,$$

for any fixed $v \in \mathbb{R}$, including $v = 0$ because $\Phi(s)$ is holomorphic and unitary on $\operatorname{Re} s = 1/2$ (use (6.28)) and real on $\mathbb{R}$.

Theorem 6.11. *The Eisenstein series $E_{\mathfrak{a}}(z, s)$ has no poles on the line $\operatorname{Re} s = 1/2$.*

Proof. Suppose $s_0 = 1/2 + iv$ is a pole of $E_{\mathfrak{a}}(z, s)$ of order $m \geq 1$, say. Since $\Phi(s)$ is regular at $s = s_0$, we have

$$u(z) = \lim_{s \to s_0} (s - s_0)^m E_{\mathfrak{a}}(z, s) = \lim_{s \to s_0} (s - s_0)^m E_{\mathfrak{a}}^Y(z, s).$$

Hence, it follows by (6.32) that $\|u\| = 0$, so $u(z) \equiv 0$ because $u(z)$ is continuous (in fact, real-analytic).

Proposition 6.12. *For $s \neq 1/2$ the Eisenstein series $E_{\mathfrak{a}}(z, s)$ does not vanish identically.*

Proof. The zero-th term of $E_{\mathfrak{a}}(\sigma_{\mathfrak{a}} z, s)$ is equal to $y^s + \varphi_{\mathfrak{a}\mathfrak{a}}(s)\, y^{1-s} \not\equiv 0$.

Show that the Eisenstein series $E_{\mathfrak{a}}(z, 1/2)$ vanishes identically if and only if $\varphi_{\mathfrak{a}\mathfrak{a}}(1/2) = -1$.

Proposition 6.13. *The point $s = 1$ is a pole of $E_{\mathfrak{a}}(z, s)$ with residue*

$$(6.33) \qquad \operatorname*{res}_{s=1} E_{\mathfrak{a}}(z, s) = |F|^{-1}.$$

Proof. Suppose $\varphi_{\mathfrak{a}\mathfrak{a}}(s)$ is regular at $s = \sigma > 1/2$. Letting $v \to 0$ in (6.31), we get

$$\|E_{\mathfrak{a}}^Y(\sigma)\|^2 + (2\sigma - 1)^{-1} Y^{1-2\sigma} \sum_{\mathfrak{b}} |\varphi_{\mathfrak{a}\mathfrak{b}}(\sigma)|^2$$
$$= (2\sigma - 1)^{-1} Y^{2\sigma-1} + 2\, \varphi_{\mathfrak{a}\mathfrak{a}}(\sigma) \log Y - \varphi'_{\mathfrak{a}\mathfrak{a}}(\sigma).$$

Hence,

$$\lim_{\sigma \to 1} (\sigma - 1)^2 \|E_{\mathfrak{a}}^Y(\sigma)\|^2 = \alpha + O(Y^{-1}),$$

where α is the residue of $\varphi_{\mathfrak{a}\mathfrak{a}}(s)$ at $s = 1$. On the other hand, the residue of $E_{\mathfrak{a}}(z, s)$ at $s = 1$ is an eigenfunction of Δ with eigenvalue zero, so it is a harmonic function in $\mathfrak{L}(\Gamma \backslash \mathbb{H})$; hence, it is constant. By the Fourier expansion this constant is equal to the residue of $\varphi_{\mathfrak{a}\mathfrak{a}}(s)$, whereas the other coefficients must be regular at $s = 1$. Therefore,

$$\lim_{s \to 1} (s - 1)\, E_{\mathfrak{a}}(z, s) = \alpha.$$

Comparing the two limits, we infer that $\alpha^2 |F| = \alpha + O(Y^{-1})$. Letting $Y \to +\infty$, we obtain $\alpha = |F|^{-1}$, as claimed.

Now we provide a proof of the Maass-Selberg relations for arbitrary Maass forms which do not grow exponentially at the cusps. Such forms have the Fourier expansion (3.4). As with the Eisenstein series, we truncate $f(z)$ by subtracting the zero-th terms in cuspidal zones, *i.e.* we put

$$
f^Y(z) = \begin{cases} f(z) - f_{\mathfrak{a}}(\operatorname{Im}\sigma_{\mathfrak{a}}^{-1}z) & \text{if } z \in F_{\mathfrak{a}}(Y), \\ f(z) & \text{if } z \in F(Y). \end{cases}
$$

By (3.7) the truncated form has exponential decay at cusps, *i.e.*

$$
f^Y(z) \ll e^{-2\pi y(z)}, \qquad \text{for } z \in F.
$$

Theorem 6.14. *Let* $f \in \mathcal{A}_{s_1}(\Gamma\backslash\mathbb{H})$ *and* $g \in \mathcal{A}_{s_2}(\Gamma\backslash\mathbb{H})$. *Suppose* f, g *satisfy* (3.3) *at any cusp. Then, for* Y *sufficiently large, we have*

$$
(6.34) \qquad (\lambda_1 - \lambda_2)\langle f^Y, \overline{g^Y}\rangle = \sum_{\mathfrak{a}} \left(f_{\mathfrak{a}}(Y)\, g_{\mathfrak{a}}'(Y) - f_{\mathfrak{a}}'(Y)\, g_{\mathfrak{a}}(Y) \right),
$$

where $\lambda_1 = s_1(1 - s_1)$ *and* $\lambda_2 = s_2(1 - s_2)$.

Proof. We begin by applying Green's formula (the hyperbolic version):

$$
(\lambda_1 - \lambda_2)\int_{F(Y)} fg\, d\mu = \int_{F(Y)} (f\,\Delta g - g\,\Delta f)\, d\mu
$$
$$
= \int_{\partial F(Y)} \left(f\,\frac{\partial g}{\partial \mathbf{n}} - g\,\frac{\partial f}{\partial \mathbf{n}} \right) d\ell.
$$

The boundary $\partial F(Y)$ consists of segments of sides of F and the horocycles $\sigma_{\mathfrak{a}}L(Y)$, where $L(Y) = \{z = x+iY : 0 < x < 1\}$ (the beginning of cuspidal zones, see (2.4)). Since the integrals along the segments of equivalent sides cancel out, we are left with

$$
(\lambda_1 - \lambda_2)\int_{F(Y)} fg\, d\mu
$$
$$
= \sum_{\mathfrak{a}} \int_{L(Y)} \left(f(\sigma_{\mathfrak{a}}z)\,\frac{\partial}{\partial y}g(\sigma_{\mathfrak{a}}z) - g(\sigma_{\mathfrak{a}}z)\,\frac{\partial}{\partial y}f(\sigma_{\mathfrak{a}}z) \right) dx
$$

after the change of variable $z \mapsto \sigma_{\mathfrak{a}}z$. Next, by the Fourier expansion

$$
f(\sigma_{\mathfrak{a}}z) = \sum_n f_n(y)\, e(nx) \qquad \text{with } f_0(y) = f_{\mathfrak{a}}(y),
$$
$$
g(\sigma_{\mathfrak{a}}z) = \sum_n g_n(y)\, e(nx) \qquad \text{with } g_0(y) = g_{\mathfrak{a}}(y),
$$

we get

$$
\int_{L(Y)} = \sum_n \left(f_n(Y)\, g_{-n}'(Y) - f_n'(Y)\, g_{-n}(Y) \right)
$$

after integration in $0 < x < 1$. Furthermore, by the Whittaker differential equation (1.25) for the Fourier coefficients,

$$\frac{d}{dy}\big(f_n(y)\, g'_{-n}(y) - f'_n(y)\, g_{-n}(y)\big) = (\lambda_1 - \lambda_2)\, y^{-2}\, f_n(y)\, g_{-n}(y).$$

If $n \neq 0$ this has an exponential decay as $y \to +\infty$, so we can integrate in $y > Y$, getting

$$f_n(Y)\, g'_{-n}(Y) - f'_n(Y)\, g_{-n}(Y) = (\lambda_2 - \lambda_1) \int_Y^{+\infty} f_n(y)\, g_{-n}(y)\, y^{-2}\, dy.$$

Summing over $n \neq 0$, we infer that

$$\int_{L(Y)} = f_{\mathfrak{a}}(Y)\, g'_{\mathfrak{a}}(Y) - f'_{\mathfrak{a}}(Y)\, g_{\mathfrak{a}}(Y)$$
$$+ (\lambda_2 - \lambda_1) \int_Y^{+\infty} \int_0^1 f^Y(\sigma_{\mathfrak{a}} z)\, g^Y(\sigma_{\mathfrak{a}} z)\, d\mu z.$$

Here the last integral is equal to (after the change $z \mapsto \sigma_{\mathfrak{a}}^{-1} z$)

$$\int_{F_{\mathfrak{a}}(Y)} f^Y(z)\, g^Y(z)\, d\mu z.$$

Finally, summing over the cusps, we arrive at (6.34) by collecting these integrals together with the one we began with.

Remarks. If the zero-th terms are of type $f_{\mathfrak{a}}(y) = f_{\mathfrak{a}}^+ \, y^s + f_{\mathfrak{a}}^- \, y^{1-s}$, then (6.34) becomes

$$\langle f^Y, \overline{g^Y} \rangle = (s_1 - s_2)^{-1} \sum_{\mathfrak{a}} \big(f_{\mathfrak{a}}^+ \, g_{\mathfrak{a}}^- \, Y^{s_1 - s_2} - f_{\mathfrak{a}}^- \, g_{\mathfrak{a}}^+ \, Y^{s_1 - s_2}\big)$$
$$+ (s_1 + s_2 - 1)^{-1} \sum_{\mathfrak{a}} \big(f_{\mathfrak{a}}^+ \, g_{\mathfrak{a}}^+ \, Y^{s_1 + s_2 - 1} - f_{\mathfrak{a}}^- \, g_{\mathfrak{a}}^- \, Y^{1 - s_1 - s_2}\big)$$

upon dividing by $\lambda_1 - \lambda_2 = (s_1 - s_2)(1 - s_1 - s_2)$, which requires the condition $\lambda_1 \neq \lambda_2$. In particular, for the Eisenstein series this reduces to (6.30).

Now that we know that the Eisenstein series are holomorphic on the critical line $\mathrm{Re}\, s = 1/2$, we can extend (6.30) by examining carefully what happens at $s_1 = \bar{s}_2 = \sigma + iv$ as $\sigma \to 1/2$. In matrix notation all the relations (6.30) read simultaneously as the following one:

$$\langle \mathcal{E}^Y(\cdot, s), {}^t\mathcal{E}^Y(\cdot, s) \rangle = (2iv)^{-1}\big(\Phi(\bar{s})\, Y^{2iv} - \Phi(s)\, Y^{-2iv}\big)$$
$$+ (2\sigma - 1)^{-1}\big(Y^{2\sigma - 1} - \Phi(s)\, \Phi(\bar{s}) Y^{1 - 2\sigma}\big),$$

where $s = \sigma + iv$, $v \neq 0$. Here we apply the following approximations

$$Y^{2\sigma-1} = 1 + (2\sigma - 1)\log Y + \cdots,$$

$$Y^{1-2\sigma} = 1 - (2\sigma - 1)\log Y + \cdots,$$

$$\Phi(\sigma + iv) = \Phi(s) + (\sigma - 1/2)\,\Phi'(s) + \cdots,$$

$$\Phi(\sigma + iv)\,\Phi(\sigma - iv) = 1 + (2\sigma - 1)\,\Phi'(s)\,\Phi(s)^{-1} + \cdots,$$

where $s = 1/2 + iv$ (note that for the last approximation one needs the functional equation (6.23)). Hence, upon taking the limit $\sigma \to 1/2$ we derive

$$\langle \mathcal{E}^Y(\cdot, s), {}^t\mathcal{E}^Y(\cdot, s)\rangle = (2s - 1)^{-1}\big(\Phi(1 - s)\,Y^{2s-1} - \Phi(s)\,Y^{1-2s}\big)$$

$$\text{(6.35)} \qquad\qquad + 2\log Y - \Phi'(s)\,\Phi(s)^{-1},$$

for $\operatorname{Re} s = 1/2$, $s \neq 1/2$. Furthermore, at the center of the critical strip

$$\text{(6.36)} \qquad \langle \mathcal{E}^Y(\cdot, 1/2), {}^t\mathcal{E}^Y(\cdot, 1/2)\rangle = (2\log Y - \Phi'(1/2))\,(1 + \Phi(1/2))\,.$$

Chapter 7

The Spectral Theorem. Continuous Part

To complete the decomposition of the space $\mathfrak{L}(\Gamma\backslash\mathbb{H})$ into Δ-inva- riant subspaces it remains to do it in the subspace $\mathcal{E}(\Gamma\backslash\mathbb{H})$ spanned by the incomplete Eisenstein series $E_\mathfrak{a}(z|\psi)$ (the orthogonal complement $\mathcal{C}(\Gamma\backslash\mathbb{H})$ was already shown to be spanned by Maass cusp forms, see Chapter 4). The spectral expansion for the incomplete Eisenstein series

$$(7.1) \qquad E_\mathfrak{a}(z|\psi) = \sum_{\gamma\in\Gamma_\mathfrak{a}\backslash\Gamma} \psi(\operatorname{Im}\sigma_\mathfrak{a}^{-1}\gamma z)$$

with $\psi \in C_0^\infty(\mathbb{R}^+)$ will emerge from the contour integral (see (3.12))

$$(7.2) \qquad E_\mathfrak{a}(z|\psi) = \frac{1}{2\pi i}\int\limits_{(\sigma)} \hat{\psi}(s)\, E_\mathfrak{a}(z,s)\, ds$$

after moving the integration from $\operatorname{Re} s = \sigma > 1$ to the line $\sigma = 1/2$ followed by an application of the functional equation for the Eisenstein series. Recall the bound (which is uniform in vertical strips)

$$(7.3) \qquad \hat{\psi}(s) = \int_0^{+\infty} \psi(y)\, y^{-s-1}\, dy \ll (|s|+1)^{-A}\,.$$

This approach requires some control over the growth of $E_\mathfrak{a}(z,s)$ in the s variable. So far our knowledge is rather poor in this aspect, namely that $E_\mathfrak{a}(z,s)$ is a meromorphic function in s of order ≤ 8. We need a polynomial growth on average over segments of the line $\operatorname{Re} s = 1/2$.

7.1. The Eisenstein transform.

Consider the subspace $C_0^\infty(\mathbb{R}^+)$ of the Hilbert space $\mathfrak{L}^2(\mathbb{R}^+)$ with the inner product

$$(7.4) \qquad\qquad \langle f, g \rangle = \frac{1}{2\pi} \int_0^{+\infty} f(r)\, \overline{g}(r)\, dr \,.$$

To a cusp $\mathfrak{a}$ we associate the Eisenstein transform

$$E_\mathfrak{a} : C_0^\infty(\mathbb{R}^+) \longrightarrow \mathcal{A}(\Gamma \backslash \mathbb{H})$$

defined by

$$(7.5) \qquad\qquad (E_\mathfrak{a} f)(z) = \frac{1}{4\pi} \int_0^{+\infty} \overset{\cdot}{f}(r)\, E_\mathfrak{a}(z, 1/2 + ir)\, dr \,.$$

The estimate (6.20) shows that the Eisenstein series $E_\mathfrak{a}(z, 1/2 + ir)$ barely fails to be square-integrable on F. However, by partial integration in r we get a slightly better bound for the Eisenstein transform, namely

$$(7.6) \qquad\qquad (E_\mathfrak{a} f)(\sigma_\mathfrak{b} z) \ll y^{1/2}(\log y)^{-1} \qquad \text{as } y \to +\infty$$

at any cusp $\mathfrak{b}$. The gain of the logarithmic factor is small, yet sufficient to see that the Eisenstein transform is in $\mathfrak{L}(\Gamma \backslash \mathbb{H})$, *i.e.*

$$E_\mathfrak{a} : C_0^\infty(\mathbb{R}^+) \longrightarrow \mathfrak{L}(\Gamma \backslash \mathbb{H}) \,.$$

Proposition 7.1. *For $f, g \in C_0^\infty(\mathbb{R})$ we have*

$$(7.7) \qquad\qquad \langle E_\mathfrak{a} f, E_\mathfrak{b} g \rangle = \delta_{\mathfrak{a}\mathfrak{b}} \langle f, g \rangle \,.$$

Proof. For the proof we consider the truncated Eisenstein transform

$$(E_\mathfrak{a}^Y f)(z) = \frac{1}{4\pi} \int_0^{+\infty} f(r)\, E_\mathfrak{a}^Y(z, 1/2 + ir)\, dr,$$

where $E_\mathfrak{a}^Y(z, s)$ is the truncated Eisenstein series (see (6.29)). We get the approximation

$$(E_\mathfrak{a}^Y f)(z) = (E_\mathfrak{a} f)(z) + O\!\left(\frac{y(z)^{1/2}}{\log y(z)} \right)$$

on integrating by parts in r, wherein the equality holds if $z \in F(Y)$. Hence, we infer that

$$\| (E_\mathfrak{a} - E_\mathfrak{a}^Y) f \| \ll (\log Y)^{-1/2} \,,$$

and by the Cauchy-Schwarz inequality this gives the approximation

$$\langle E_\mathfrak{a} f, E_\mathfrak{b} g \rangle = \langle E_\mathfrak{a}^Y f, E_\mathfrak{b}^Y g \rangle + O\big((\log Y)^{-1/2}\big).$$

Next we compute the inner product

$$\langle E_\mathfrak{a}^Y f, E_\mathfrak{b}^Y g \rangle$$
$$= \frac{1}{(4\pi)^2} \int_0^{+\infty}\!\!\int_0^{+\infty} f(r')\,\overline{g}(r)\,\langle E_\mathfrak{a}^Y(\cdot, 1/2 + ir'), E_\mathfrak{b}^Y(\cdot, 1/2 + ir) \rangle\, dr\, dr'$$

by an appeal to the Maass-Selberg formula (see Proposition 6.8)

$$\langle E_\mathfrak{a}^Y(\cdot, 1/2 + ir'), E_\mathfrak{b}^Y(\cdot, 1/2 + ir) \rangle$$
$$= \frac{i}{r' + r}\, \varphi_{\mathfrak{a}\mathfrak{b}}(1/2 + ir)\, Y^{i(r'+r)}$$
$$- \frac{i}{r' + r}\, \varphi_{\mathfrak{a}\mathfrak{b}}(1/2 + ir')\, Y^{-i(r'+r)}$$
$$+ \frac{i}{r - r'}\, \big(\delta_{\mathfrak{a}\mathfrak{b}} - \Phi_\mathfrak{a}(1/2 + ir')\, \Phi_\mathfrak{b}(1/2 + ir)\big) Y^{i(r-r')}$$
$$+ \frac{i}{r - r'}\, \delta_{\mathfrak{a}\mathfrak{b}} \big(Y^{i(r'-r)} - Y^{i(r-r')}\big).$$

Since all terms are continuous in $(r, r') \in \mathbb{R}^+ \times \mathbb{R}^+$ (recall that $\Phi(s)$ is unitary on $\operatorname{Re} s = 1/2$) we gain the factor $\log Y$ by partial integration in r for all but the last term. We obtain

$$\langle E_\mathfrak{a}^Y f, E_\mathfrak{b}^Y g \rangle$$
$$= \frac{\delta_{\mathfrak{a}\mathfrak{b}}}{(4\pi)^2} \int_0^{+\infty}\!\!\int_0^{+\infty} f(r')\,\overline{g}(r)\, \frac{Y^{i(r'-r)} - Y^{i(r-r')}}{i(r'-r)}\, dr\, dr' + O\big((\log Y)^{-1}\big).$$

Since r is bounded below by a positive constant, the innermost integral in r' approximates to

$$f(r) \int_{-\infty}^{+\infty} 2\sin(u \log Y)\, \frac{du}{u} = 2\pi\, f(r)$$

up to an error term $O((\log Y)^{-1})$, which is estimated by partial integration. Collecting the above results and letting $Y \to +\infty$, we get (7.7).

Corollary. *The Eisenstein transform $E_\mathfrak{a}$ maps $C_0^\infty(\mathbb{R}^+)$ into $\mathfrak{L}(\Gamma\backslash\mathbb{H})$ isometrically.*

Remark. One can extend the Eisenstein transform $E_\mathfrak{a}$ to an isometry of $\mathfrak{L}^2(\mathbb{R}^+)$ into $\mathfrak{L}(\Gamma\backslash\mathbb{H})$, of course not surjectively. This is a close analogue of the Plancherel theorem for the Fourier transform.

The image $\mathcal{E}_{\mathfrak{a}}(\Gamma\backslash\mathbb{H})$ of the Eisenstein transform $E_{\mathfrak{a}}$ is called *the space of the Eisenstein series $E_{\mathfrak{a}}(z,s)$.* Clearly $\mathcal{E}_{\mathfrak{a}}(\Gamma\backslash\mathbb{H})$ is an invariant subspace for the Laplace operator; more precisely, Δ acts on $\mathcal{E}_{\mathfrak{a}}(\Gamma\backslash\mathbb{H})$ through multiplication, *i.e.*

$$\Delta\, E_{\mathfrak{a}} = E_{\mathfrak{a}}\, M,$$

where

$$(Mf)(r) = -\left(r^2 + \frac{1}{4}\right) f(r).$$

There are various orthogonalities worthy of record. By Proposition 7.1 the Eisenstein spaces $\mathcal{E}_{\mathfrak{a}}(\Gamma\backslash\mathbb{H})$ for distinct cusps are orthogonal. Also every $\mathcal{E}_{\mathfrak{a}}(\Gamma\backslash\mathbb{H})$ is orthogonal to the space $\mathcal{C}(\Gamma\backslash\mathbb{H})$ spanned by cusp forms, by Theorem 6.7. Finally, $\mathcal{E}_{\mathfrak{a}}(\Gamma\backslash\mathbb{H})$ is orthogonal to the residues of any Eisenstein series $E_{\mathfrak{b}}(z,s)$ at poles $s = s_j$ in $1/2 < s_j \le 1$, because the eigenvalues satisfy $0 \le \lambda_j = s_j(1 - s_j) < 1/4 \le r^2 + 1/4$ for $r \in \mathbb{R}$. Therefore, arguing by means of orthogonality, we conclude that

$$\mathcal{R}(\Gamma\backslash\mathbb{H}) \underset{\mathfrak{a}}{\oplus} \mathcal{E}_{\mathfrak{a}}(\Gamma\backslash\mathbb{H}) \subset \mathcal{E}(\Gamma\backslash\mathbb{H}),$$

where $\mathcal{R}(\Gamma\backslash\mathbb{H})$ is the space spanned by residues of all the Eisenstein series in the segment $(1/2, 1]$. The spectral theorem will show that $\mathcal{R}(\Gamma\backslash\mathbb{H})$ together with $\mathcal{E}_{\mathfrak{a}}(\Gamma\backslash\mathbb{H})$ densely fill the space $\mathcal{E}(\Gamma\backslash\mathbb{H})$.

7.2. Bessel's inequality.

Suppose the f_j are mutually orthogonal in a Hilbert space. Then for any f in that space we have

$$\|f - \sum_j f_j\|^2 = \|f\|^2 - 2\,\mathrm{Re}\,\sum_j \langle f, f_j\rangle + \sum_j \|f_j\|^2.$$

In particular, if the f_j are chosen so that $\langle f, f_j\rangle = \|f_j\|^2$, this gives

$$\|f - \sum_j f_j\|^2 = \|f\|^2 - \sum_j \|f_j\|^2,$$

and hence, by the positivity of the norm $\|\cdot\|$, we obtain the Bessel inequality

$$(7.8) \qquad\qquad \sum_j \|f_j\|^2 \le \|f\|^2$$

(to be precise, use the above relations for a finite collection of functions f_j and then drop this condition in (7.8) by positivity).

We employ Bessel's inequality in the space $\mathfrak{L}(\Gamma\backslash\mathbb{H})$ for an automorphic kernel

$$f(z) = K(z, w) = \sum_{\gamma \in \Gamma} k(z, \gamma w),$$

where z is the variable and w is fixed. Suppose $k(u)$ is smooth and compactly supported on $\mathbb{R}^+$; then $f(z)$ is also compactly supported on $\mathbb{H}$, so it belongs to $\mathfrak{L}(\Gamma\backslash\mathbb{H})$. Anticipating the spectral expansion for $K(z, w)$ (see Theorem 8.1), we choose the functions

$$f_j(z) = h(t_j) \, u_j(z) \, \overline{u_j}(w) \,,$$

$$f_{\mathfrak{a}}(z) = \frac{1}{4\pi} \int_A^B h(r) \, E_{\mathfrak{a}}(z, 1/2 + ir) \, \overline{E_{\mathfrak{a}}}(w, 1/2 + ir) \, dr$$

to approximate $f(z)$, so that the inequality (7.8) is quite sharp. Here the $u_j(z)$ range over an orthogonal system in the space of discrete spectrum (more precisely, at this moment we can only take $u_j(z)$ in the space of Maass cusp forms and the residues of the Eisenstein series in the segment $(1/2, 1]$), the $E_{\mathfrak{a}}(z, s)$ are the Eisenstein series and $h(r)$ is the Selberg/Harish-Chandra transform of $k(u)$. The integral is cut off at fixed heights A, B with $0 < A < B < +\infty$, because we do not know yet if the full integral converges. Observe that $f_{\mathfrak{a}}(z)$ is the Eisenstein transform

$$f_{\mathfrak{a}}(z) = (E_{\mathfrak{a}}g)(z)$$

for g given by

$$g(r) = h(r) \, \overline{E_{\mathfrak{a}}}(w, 1/2 + ir)$$

if $A \leq r \leq B$, and $g(r) = 0$ elsewhere. This is not smooth at the end points; nevertheless Proposition 7.1 remains valid by a suitable approximation or by essentially the same proof.

From the discussion concluding the previous section we have learned that all the $f_j(z)$ and $f_{\mathfrak{a}}(z)$ are mutually orthogonal. To compute the projections of the f_j on the kernel $f(z) = K(z, w)$ we appeal to Theorem 1.16. We get

$$\langle f, f_j \rangle = \overline{h}(t_j) \, u_j(w) \, \langle f, u_j \rangle$$

$$= \overline{h}(t_j) \, u_j(w) \int_{\mathbb{H}} k(z, w) \, \overline{u}_j(z) \, d\mu z$$

$$= |h(t_j) \, u_j(w)|^2 = \|f_j\|^2 \,.$$

Similarly, by Proposition 7.1, we get

$$\langle f, f_{\mathfrak{a}} \rangle = \frac{1}{2\pi} \int_A^B |h(r) \, E_{\mathfrak{a}}(w, 1/2 + ir)|^2 \, dr$$

$$= \frac{1}{2\pi} \int_A^B |g(r)|^2 \, dr = \|f_{\mathfrak{a}}\|^2 \,.$$

Now all the conditions for Bessel's inequality are satisfied; hence

(7.9)
$$\sum_j |h(t_j)\, u_j(w)|^2 + \sum_{\mathfrak{a}} \frac{1}{4\pi} \int_{-\infty}^{+\infty} |h(r)\, E_{\mathfrak{a}}(w, 1/2 + ir)|^2\, dr$$
$$\leq \int_F |K(z, w)|^2\, d\mu z\,.$$

Here we have dropped the restriction $A \leq r \leq B$ by positivity, and we have added integrals over negative r by symmetry.

A particularly useful inequality comes out of (7.9) for the kernel $k(u)$ which is the characteristic function of the segment $0 \leq u \leq \delta$ with small δ, say $\delta \leq 1/4$ (this is an admissible kernel, but if you feel uncomfortable with the discontinuity think of $k(u)$ as a compactly supported smooth approximation to this characteristic function).

First we estimate the Selberg/Harish-Chandra transform $h(t)$. To this end, rather than computing explicitly, we appeal to the integral representation

$$h(t) = \int_{\mathbb{H}} k(i, z)\, y^s\, d\mu z$$

as in the proof of Theorem 1.16. For $s = 0$ this gives

$$h\Big(\frac{i}{2}\Big) = \int_{\mathbb{H}} k(i, z)\, d\mu z = 4\pi\delta\,,$$

which is just the hyperbolic area of a disc of radius r given by $\sinh(r/2) = \sqrt{\delta}$. Since $u(i, z) < \delta$ implies $|y - 1| < 4\sqrt{\delta}$, we have $|y^s - 1| \leq |s|\, |y - 1| \leq 4|s|\sqrt{\delta}$, whence

$$|h(t) - h\Big(\frac{i}{2}\Big)| \leq 4|s|\sqrt{\delta}\, h\Big(\frac{i}{2}\Big)\,.$$

This yields $2\pi\delta < |h(t)| < 6\pi\delta$, if $|s| \leq (8\sqrt{\delta})^{-1}$.

Next we estimate the $\mathfrak{L}^2$-norm of $K(z, w)$. We begin by

$$\int_F |K(z, w)|^2\, d\mu z = \sum_{\gamma, \gamma' \in \Gamma} \int_F k(\gamma' z, w)\, k(\gamma' z, \gamma w)\, d\mu z$$
$$= \sum_{\gamma \in \Gamma} \int_{\mathbb{H}} k(z, w)\, k(z, \gamma w)\, d\mu z\,.$$

Here we have $u(z, w) \leq \delta$ and $u(z, \gamma w) \leq \delta$; hence $u(\omega, \gamma w) \leq 4\delta(\delta + 1)$ by the triangle inequality for the hyperbolic distance. Setting

$$N_\delta(w) = \#\{\gamma \in \Gamma :\ u(w, \gamma w) \leq 4\delta(\delta + 1)\}\,,$$

we obtain

$$\int_F |K(z,w)|^2 \, d\mu z \le N_\delta(w) \int_H k(i,z)^2 \, d\mu z = N_\delta(w) \, h\left(\frac{i}{2}\right).$$

Inserting the above estimates into (7.9), we get

$$\sum_j' |u_j(z)|^2 + \sum_{\mathfrak{a}} \frac{1}{4\pi} \int' |E_{\mathfrak{a}}(z, 1/2 + ir)|^2 \, dr < (\pi \delta)^{-1} N_\delta(z),$$

where $'$ restricts the summation and the integration to points $s_j = 1/2 + it_j$ and $s = 1/2 + ir$ in the disc $|s| \le (8\sqrt{\delta})^{-1}$. By Corollary 2.12 we get $N_\delta(z) \ll \sqrt{\delta} \, y_\Gamma(z) + 1$. Choosing $\delta = (4T)^{-2}$, we obtain

Proposition 7.2. *Let $T \ge 1$ and $z \in \mathbb{H}$. We have*

$$(7.10) \qquad \sum_{|t_j| < T} |u_j(z)|^2 + \sum_{\mathfrak{a}} \int_{-T}^{T} |E_{\mathfrak{a}}(z, 1/2 + it)|^2 \, dt \ll T^2 + T \, y_\Gamma(z),$$

where $y_\Gamma(z)$ is the invariant height of z (see (2.42)) and the implied constant depends on the group Γ alone.

One can derive from (7.10) many valuable estimates for the spectra and the eigenfunctions. For example, one can show quickly that

$$(7.11) \qquad N_\Gamma(T) = \#\{j : \ |t_j| < T\} \ll T^2.$$

To this end, integrate (7.10) over the central part $F(Y) \subset F$ with $Y \asymp T$ and use Theorem 3.2 to extend the integral of $|u_j(z)|^2$ over the whole fundamental domain at the cost of a small error term. Ignoring the integrals of $|E_{\mathfrak{a}}(z, 1/2 + it)|^2$, one obtains (7.11). For the continuous spectrum analogue of this result, see (10.13).

7.3. Spectral decomposition of $\mathcal{E}(\Gamma \backslash \mathbb{H})$.

Since $\mathcal{E}(\Gamma \backslash \mathbb{H})$ is spanned by the incomplete Eisenstein series $E_{\mathfrak{a}}(z|\psi)$, it suffices to decompose $E_{\mathfrak{a}}(z|\psi)$ for any $\psi \in C_0^\infty(\mathbb{R}^+)$. The Mellin transform of ψ is entire, and it satisfies the bound (7.3) in any vertical strip. Thus the integral (7.2) converges absolutely if $\sigma > 1$. Moving to the line $\operatorname{Re} s = 1/2$ (which is justified because $E_{\mathfrak{a}}(z,s)$ has at most polynomial growth in s on average due to Proposition 7.2 and the Phragmén-Lindelöf convexity principle), we pass a finite number of simple poles in the segment $(1/2, 1]$ and get

$$(7.12) \qquad E_{\mathfrak{a}}(z|\psi) = \sum_{1/2 < s_j \le 1} \hat{\psi}(s_j) \, u_{\mathfrak{a}j}(z) + \frac{1}{2\pi i} \int_{(1/2)} \hat{\psi}(s) \, E_{\mathfrak{a}}(z,s) \, ds,$$

where $u_{\mathfrak{a}j}(z)$ is the residue of $E_{\mathfrak{a}}(z, s)$ at $s = s_j$. Here the coefficients $\hat{\psi}(s_j)$ are given by the inner product

$$(7.13) \qquad \hat{\psi}(s_j) = \langle E_{\mathfrak{a}}(\cdot|\psi), u_{\mathfrak{a}j} \rangle \, \|u_{\mathfrak{a}j}\|^{-2},$$

because the $u_{\mathfrak{a}j}(z)$ are mutually orthogonal as well as being orthogonal to each of the Eisenstein series $E_{\mathfrak{a}}(z, s)$ on the line $\operatorname{Re} s = 1/2$.

The above argument, however, does not apply to $\hat{\psi}(s)$ on the line $\operatorname{Re} s = 1/2$, because the inner product $\langle E_{\mathfrak{a}}(\cdot, s), E_{\mathfrak{a}}(\cdot, s') \rangle$ diverges. Therefore the expansion (7.12) cannot be regarded as a spectral decomposition (in the sense of a continuous eigenpacket), since the coefficient $\hat{\psi}(s)$ in the integral does not agree with the projection of $E_{\mathfrak{a}}(\cdot|\psi)$ on $E_{\mathfrak{a}}(\cdot, s)$. To get the proper representation we rearrange this integral by an appeal to the functional equation

$$E_{\mathfrak{a}}(z, 1 - s) = \sum_{\mathfrak{b}} \varphi_{\mathfrak{a}\mathfrak{b}}(1 - s) \, E_{\mathfrak{b}}(z, s)$$

(see (6.23)). We also use the formula (see Lemma 3.2)

$$\langle E_{\mathfrak{a}}(\cdot|\psi), E_{\mathfrak{b}}(\cdot, s) \rangle = \int_0^{+\infty} \left(\delta_{\mathfrak{a}\mathfrak{b}} \, y^{1-s} + \varphi_{\mathfrak{a}\mathfrak{b}}(1 - s) \, y^s \right) \psi(y) \, y^{-2} \, dy.$$

Multiplying this by $E_{\mathfrak{b}}(z, s)$ and summing over $\mathfrak{b}$, we get

$$\sum_{\mathfrak{b}} \langle E_{\mathfrak{a}}(\cdot|\psi), E_{\mathfrak{b}}(\cdot, s) \rangle \, E_{\mathfrak{b}}(z, s) = \hat{\psi}(s) \, E_{\mathfrak{a}}(z, s) + \hat{\psi}(1 - s) \, E_{\mathfrak{a}}(z, 1 - s)$$

by the functional equation. Finally, integrating this in s on the line $\operatorname{Re} s = 1/2$, we obtain

$$
(7.14) \qquad
\begin{aligned}
&\frac{1}{2\pi i} \int_{(1/2)} \hat{\psi}(s) \, E_{\mathfrak{a}}(z, s) \, ds \\
&\qquad = \sum_{\mathfrak{b}} \frac{1}{4\pi i} \int_{(1/2)} \langle E_{\mathfrak{a}}(\cdot|\psi), E_{\mathfrak{b}}(\cdot, s) \rangle \, E_{\mathfrak{b}}(z, s) \, ds.
\end{aligned}
$$

This is the desired expression for the projection on the Eisenstein series. Note that it takes all the Eisenstein series $E_{\mathfrak{b}}(z, 1/2 + ir)$ to perform the spectral decomposition of one $E_{\mathfrak{a}}(z|\psi)$.

The expansions (7.12)–(7.14) extend to all functions $f \in \mathcal{E}(\Gamma\backslash\mathbb{H})$ by linearity. Some of the residues $u_{\mathfrak{a}j}(z)$ of the Eisenstein series $E_{\mathfrak{a}}(z, s)$ associated with distinct cusps can be linearly dependent—for instance, the residues at $s = 1$ are all equal to a constant (see (6.33)). Moreover, the residues need to be normalized so as to give the $\mathfrak{L}^2$-norm one. We let $\mathcal{R}_{s_j}(\Gamma\backslash\mathbb{H})$ be the space spanned by the residues of all Eisenstein series at $s = s_j$; thus

$$\dim \mathcal{R}_{s_j}(\Gamma\backslash\mathbb{H}) \leq h = \text{ the number of cusps.}$$

Then we let $\mathcal{R}(\Gamma\backslash\mathbb{H})$ be the space spanned by all residues of all Eisenstein series in the segment $(1/2, 1]$. This space has the orthogonal decomposition

$$\mathcal{R}(\Gamma\backslash\mathbb{H}) = \bigoplus_{1/2 < s_j \leq 1} \mathcal{R}_{s_j}(\Gamma\backslash\mathbb{H}).$$

In each subspace $\mathcal{R}_{s_j}(\Gamma\backslash\mathbb{H})$ we choose an orthonormal basis out of which we assemble the basis $\{u_j(z)\}$ of $\mathcal{R}(\Gamma\backslash\mathbb{H})$.

With regard to the integrals of Eisenstein series on the line $s = 1/2 + ir$, neither further rearrangement nor any normalization is desired. The collection of these Eisenstein series is called the *eigenpacket*. From the above considerations we deduce the following spectral decomposition of $\mathcal{E}(\Gamma\backslash\mathbb{H})$.

Theorem 7.3. *The space $\mathcal{E}(\Gamma\backslash\mathbb{H})$ of incomplete Eisenstein series splits orthogonally into Δ-invariant subspaces,*

$$\mathcal{E}(\Gamma\backslash\mathbb{H}) = \mathcal{R}(\Gamma\backslash\mathbb{H}) \underset{\mathfrak{a}}{\oplus} \mathcal{E}_{\mathfrak{a}}(\Gamma\backslash\mathbb{H}).$$

The spectrum of Δ in $\mathcal{R}(\Gamma\backslash\mathbb{H})$ is discrete; it consists of a finite number of points λ_j with $0 \leq \lambda_j < 1/4$. The spectrum of Δ in $\mathcal{E}_{\mathfrak{a}}(\Gamma\backslash\mathbb{H})$ is absolutely continuous; it covers the segment $[1/4, \infty)$ uniformly with multiplicity 1. Every $f \in \mathcal{E}(\Gamma\backslash\mathbb{H})$ has the expansion

(7.15)
$$f(z) = \sum_j \langle f, u_j \rangle \, u_j(z)$$
$$+ \sum_{\mathfrak{a}} \frac{1}{4\pi} \int_{-\infty}^{+\infty} \langle f, E_{\mathfrak{a}}(\cdot, 1/2 + ir) \rangle \, E_{\mathfrak{a}}(z, 1/2 + ir) \, dr,$$

which converges in the norm topology, and if f belongs to the initial domain

$$\mathcal{D}(\Gamma\backslash\mathbb{H}) = \big\{ f \in \mathcal{A}(\Gamma\backslash\mathbb{H}) : \ f, \ \Delta f \text{ smooth and bounded} \big\},$$

the expansion (7.15) converges pointwise absolutely and uniformly on compacta.

Combining Theorems 4.7 and 7.3, one gets the spectral decomposition of the whole space $\mathfrak{L}(\Gamma\backslash\mathbb{H})$. Any $f \in \mathcal{D}(\Gamma\backslash\mathbb{H})$ has the spectral expansion obtained by synthesis of (4.15) and (7.15).

7.4. Spectral expansion of automorphic kernels.

The spectral theorem is a powerful tool for analytic studies in automorphic forms. Particularly handy is the spectral series expansion for the automorphic kernel,

$$K(z,w) = \sum_{\gamma \in \Gamma} k(z, \gamma w),$$

as well as for the automorphic Green function,

$$(7.16) \qquad G_s(z/w) = \sum_{\gamma \in \Gamma} G_s(z, \gamma w).$$

First suppose $k(u) \in C_0^\infty(\mathbb{R}^+)$, so as a function of z for w on compacta this $K(z,w)$ has the absolutely and uniformly convergent spectral expansion given by (4.15) and (7.15). The projections of $K(z,w)$ on the eigenfunctions are computed in Theorem 1.16; they are

$$\langle K(\cdot, w), u_j \rangle = h(t_j)\, \overline{u}_j(w),$$
$$\langle K(\cdot, w), E_{\mathfrak{b}}(\cdot, 1/2 + ir) \rangle = h(r)\, \overline{E}_{\mathfrak{b}}(w, 1/2 + ir).$$

Hence, we obtain

Theorem 7.4. *Let $K(z,w)$ be an automorphic kernel given by a point-pair invariant $k(z,w) = k(u(z,w))$ whose Selberg/Harish-Chandra transform $h(r)$ satisfies the conditions (1.63). Then it has the spectral expansion*

$$
\begin{aligned}
K(z,w) = &\sum_j h(t_j)\, u_j(z)\, \overline{u}_j(w) \\
(7.17) \qquad &+ \sum_{\mathfrak{a}} \frac{1}{4\pi} \int_{-\infty}^{+\infty} h(r)\, E_{\mathfrak{a}}(z, 1/2 + ir)\, \overline{E}_{\mathfrak{a}}(w, 1/2 + ir)\, dr,
\end{aligned}
$$

which converges absolutely and uniformly on compacta.

Remark. Our initial assumption that $k(u)$ is compactly supported was replaced by the weaker conditions (1.63) using a suitable approximation.

Next we develop an expansion for the automorphic Green function. Formally speaking, it is a special case of an automorphic kernel for $k(u) = G_s(u)$, but the above result does not apply directly because $G_s(u)$ is singular at $u = 0$. We annihilate the singularity by considering the difference $k(u) = G_s(u) - G_a(u)$ with $\operatorname{Re} s > 1$ and $a > 1$. As a function of w the difference $G_s(z/w) - G_a(z/w)$ has the spectral expansion

$$G_s(z/w) - G_a(z/w) = \sum_j g_j(z)\, \overline{u}_j(w) + \text{ Eisenstein part.}$$

To compute the coefficients $g_j(z)$ we save work by appealing to properties of the resolvent. We get

$$g_j(z) = \int_F \left(G_s(z/w) - G_a(z/w) \right) u_j(w) \, d\mu w$$
$$= (\Delta + s(1-s))^{-1} u_j(z) - ((\Delta + a(1-a))^{-1} u_j(z)$$
$$= \chi_{sa}(s_j) \, u_j(z),$$

where

$$\chi_{sa}(v) = (s-v)^{-1}(1-s-v)^{-1} - (a-v)^{-1}(1-a-v)^{-1}.$$

Furthermore, we perform the same computations for projections on the Eisenstein series. We obtain

Theorem 7.5.. *Let* $a > 1$ *and* $\operatorname{Re} s > 1$. *Then*

$$G_s(z/w) - G_a(z/w) = \sum_j \chi_{sa}(s_j) \, u_j(z) \, \overline{u}_j(w)$$

$$(7.18) \qquad\qquad\qquad + \sum_{\mathfrak{a}} \frac{1}{4\pi i} \int_{(1/2)} \chi_{sa}(v) \, E_{\mathfrak{a}}(z,v) \, \overline{E}_{\mathfrak{a}}(w,\overline{v}) \, dv,$$

where the series and integrals converge absolutely and uniformly on compacta.

Notice that our initial domain was $\operatorname{Re} s > 1$; however, the spectral expansion (7.18) is valid in $\operatorname{Re} s > 1/2$ without modification, by analytic continuation. In order to extend the result to the complementary half-plane we consider the integral

$$I_\alpha(s) = \frac{1}{2\pi i} \int_{(\alpha)} \chi_{sa}(v) \, E_{\mathfrak{a}}(z,v) \, E_{\mathfrak{a}}(w, 1-v) \, dv,$$

where $\alpha > 1/2$ is sufficiently close to $1/2$ so that all Eisenstein series are holomorphic in the strip $1/2 < \operatorname{Re} s < \alpha$. By Cauchy's theorem,

$$I_{1/2}(s) = I_\alpha(s) - (2s-1)^{-1} E_{\mathfrak{a}}(z,s) \, E_{\mathfrak{a}}(w, 1-s)$$

for s in the strip. This furnishes the spectral expansion of $G_s(z/w) - G_a(z/w)$ in $\operatorname{Re} s < \alpha$ through the analytic continuation of $I_\alpha(s)$. Moreover, it shows the following functional equation:

$$(7.19) \qquad G_s(z/w) - G_{1-s}(z/w) = -\frac{1}{2s-1} \sum_{\mathfrak{a}} E_{\mathfrak{a}}(z,s) \, E_{\mathfrak{a}}(w, 1-s).$$

Indeed, the discrete spectrum series and the integral $I_\alpha(s)$ are invariant under the change $s \mapsto 1 - s$ within the strip $1 - \alpha < \operatorname{Re} s < \alpha$, and also

$$(7.20) \qquad \sum_{\mathfrak{a}} E_{\mathfrak{a}}(z, s)\, E_{\mathfrak{a}}(w, 1 - s) = \sum_{\mathfrak{a}} E_{\mathfrak{a}}(z, 1 - s)\, E_{\mathfrak{a}}(w, s)$$

by (6.22').

From the spectral expansion (7.18) it is plain that the point $s = s_j > 1/2$ from the discrete spectrum gives a simple pole of $G_s(z/w)$ with residue

$$\operatorname*{res}_{s=s_j} G_s(z/w) = -\frac{1}{2s_j - 1} \sum_{s_k=s_j} u_k(z)\, \overline{u}_k(w).$$

If $s = 1/2$ belongs to the discrete spectrum, then it gives a pole of order 2, but the residue comes from the Eisenstein series only. More precisely, the Laurent expansion at $s = 1/2$ begins as follows:

$$G_s(z/w) = -\frac{1}{(s - 1/2)^2} \sum_{s_j=1/2} u_j(z)\, \overline{u}_j(w)$$

$$-\frac{1}{4(s - 1/2)} \sum_{\mathfrak{a}} E_{\mathfrak{a}}(z, 1/2)\, \overline{E}_{\mathfrak{a}}(w, 1/2) + \cdots .$$

In the half-plane $\operatorname{Re} s < 1/2$ the Green function inherits a lot of simple poles from the Eisenstein series in addition to the finite number of poles at $1 - s_j$ from the discrete spectrum (see (7.19)).

Remark. The spectral theory for the resolvent operator is treated in greater generality by J. Elstrodt [El].

Chapter 8

Estimates for the Fourier Coefficients of Maass Forms

The main goal of this chapter is to establish auxiliary estimates which are needed later when checking the convergence of various series. Crude results would often do. However, since it is effortless to be general, sharp and explicit in some cases, we go beyond the primary objective.

8.1. Introduction.

Let $\{u_j(z) : j \geq 0\}$ be a complete orthonormal system of Maass forms for the discrete spectrum together with the eigenpacket $\{E_\mathfrak{c}(z, s) : s = 1/2 + it,\ t \in \mathbb{R}\}$ of Eisenstein series for the continuous spectrum for $\mathfrak{L}(\Gamma\backslash\mathbb{H})$. These have Fourier expansions of the type

$$(8.1) \qquad u_j(\sigma_\mathfrak{a} z) = \rho_{\mathfrak{a}j}(0)\, y^{1-s_j} + \sum_{n \neq 0} \rho_{\mathfrak{a}j}(n)\, W_{s_j}(nz),$$

$$(8.2) \qquad E_\mathfrak{c}(\sigma_\mathfrak{a} z, s) = \delta_{\mathfrak{a}\mathfrak{c}}\, y^s + \varphi_{\mathfrak{a}\mathfrak{c}}(s)\, y^{1-s} + \sum_{n \neq 0} \varphi_{\mathfrak{a}\mathfrak{c}}(n, s)\, W_s(nz),$$

where $\rho_{\mathfrak{a}j}(0) = 0$ if $u_j(z)$ is a cusp form or otherwise $u_j(z)$ is a linear combination of residues of Eisenstein series at $s = s_j > 1/2$. The Whittaker function satisfies $W_{s_j}(nz) \sim e(nz)$ as $n \to +\infty$; hence the tail of (8.1) looks like an expansion into exponentials. However, the terms opening the series have rather peculiar shape; they look like a power series as long as the argument is much smaller than $|t_j|$. Then the transition occurs somewhere near $2\pi |n| y \sim |t_j|$ if $|t_j|$ is large.

The coefficients $\rho_{\mathfrak{a}j}(n)$ can be exponentially large in $|t_j|$, because they must compensate the exponential decay of $W_{s_j}(nz)$ to make the norm $\|u_j\| = 1$. Actually the normalized eigenstates $u_j(z)$ are quite uniformly estimated everywhere. If u_j is a cusp form, we obtain by Theorem 3.2 that

$$(8.3) \qquad u_j(\sigma_{\mathfrak{a}}z)^2 \ll \frac{|s_j|}{y}\left(1 + \frac{1}{c_{\mathfrak{a}}y}\right),$$

where the implied constant is absolute. From this we deduce

$$(8.3') \qquad u_j(z) \ll \lambda_j^{1/4}\,(y + y^{-1})^{-1/2},$$

where the implied constant depends only on the group. For the proof we can assume that z is in a fixed fundamental polygon; see Figure 7 in Chapter 2. If z is in the central part $F(Y)$, where Y is fixed large, the result is clear. Suppose z is in the cuspidal zone $F_{\mathfrak{a}}(Y)$, so $z = \sigma_{\mathfrak{a}}w$, where w is in the semi-strip $P(Y)$, see (2.2). We have $\mathrm{Im}\,z \asymp \mathrm{Im}\,w$ if $\mathfrak{a} = \infty$, and $\mathrm{Im}\,z \asymp (\mathrm{Im}\,w)^{-1}$ if $\mathfrak{a} \neq \infty$. Hence, applying (8.3) at the point w, we obtain (8.3'). Similarly, if u_j is a residual form, then

$$(8.4) \qquad u_j(z) \ll y(z)^{1-s_j}\,.$$

For a more natural look, we scale down the Fourier coefficients to

$$(8.5) \qquad \nu_{\mathfrak{a}j}(n) = \left(\frac{4\pi|n|}{\cosh \pi t_j}\right)^{1/2} \rho_{\mathfrak{a}j}(n)\,,$$

$$(8.6) \qquad \eta_{\mathfrak{a}c}(n,t) = \left(\frac{4\pi|n|}{\cosh \pi t}\right)^{1/2} \varphi_{\mathfrak{a}c}(n, 1/2 + it)\,,$$

if $n \neq 0$. Note that $\pi(\cosh \pi t_j)^{-1} = \Gamma(s_j)\,\Gamma(1 - s_j)$ for $s_j = 1/2 + it_j$. Since of every two points $s_j, 1 - s_j$ only one counts, in subsequent writing we make a unique selection, requiring either $t_j \geq 0$ or $1/2 < s_j \leq 1$. We shall see that the coefficients so scaled are bounded on average in various ways with respect to n and t_j.

By Theorem 3.2 we infer that the Fourier coefficients of a cusp form $u_j(z)$ satisfy

$$(8.7) \qquad \sum_{0 < |n| \leq X} |\nu_{\mathfrak{a}j}(n)|^2 \ll (1 + \lambda_j^{-1})\left(|s_j| + \frac{X}{c_{\mathfrak{a}}}\right),$$

where the implied constant is absolute. Hence

$$(8.8) \qquad \nu_{\mathfrak{a}j}(n) \ll \left(|t_j| + \frac{|n|}{c_{\mathfrak{a}}}\right)^{1/2},$$

where the implied constant is absolute.

8.2. The Rankin-Selberg L-function.

A more precise estimate than (8.7), at least when X is sufficiently large, can be established using analytic properties of the series

$$(8.9) \qquad L_{\mathfrak{a}j}(s) = \sum_{n \neq 0} |\nu_{\mathfrak{a}j}(n)|^2 |n|^{-s} \,.$$

The required properties are inherited from those of the Eisenstein series $E_{\mathfrak{a}}(z, s)$ through the following integral representation:

$$(8.10) \qquad \Theta_j(s) \, L_{\mathfrak{a}j}(s) = 8 \int_{\Gamma \backslash \mathbb{H}} |u_j(z)|^2 E_{\mathfrak{a}}(z, s) \, d\mu z,$$

where

$$(8.11) \qquad \Theta_j(s) = \pi^{-1-s} \, \Gamma(s)^{-1} \Gamma(\tfrac{s}{2})^2 \, \Gamma(\tfrac{s}{2} + it_j) \, \Gamma(\tfrac{s}{2} - it_j) \cosh \pi t_j \,.$$

In the half-plane $\operatorname{Re} s > 1$, where the series (8.9) converges absolutely by virtue of (8.7), this integral representation is derived by unfolding the integral as follows:

$$\sum_{\gamma \in \Gamma_{\mathfrak{a}} \backslash \Gamma} \int_{\Gamma \backslash \mathbb{H}} |u_j(z)|^2 \, (\operatorname{Im} \sigma_{\mathfrak{a}}^{-1} \gamma z)^s \, d\mu z$$

$$= \int_{B \backslash \mathbb{H}} |u_j(\sigma_{\mathfrak{a}} z)|^2 \, y^s \, d\mu z = \sum_{n \neq 0} |\rho_{\mathfrak{a}j}(n)|^2 \int_0^{+\infty} |W_{s_j}(ny)|^2 \, y^{s-2} \, dy$$

$$= \frac{1}{\pi} \cosh(\pi t_j) \, L_{\mathfrak{a}j}(s) \, (2\pi)^{-s} \int_0^{+\infty} |K_{it_j}(y)|^2 \, y^{s-1} \, dy \,.$$

Here the Mellin transform of $|K_{it_j}(y)|^2$ is given by the product of gamma functions (see Appendix B.4), which leads us to (8.10).

From (8.10) we deduce that $L_{\mathfrak{a}j}(s)$ has an analytic continuation over the whole s-plane. In the half-plane $\operatorname{Re} s \geq 1/2$ the poles of $L_{\mathfrak{a}j}(s)$ are among those of $E_{\mathfrak{a}}(z, s)$, so they are simple and lie in the segment $(1/2, 1]$. At $s = 1$ the residue is

$$(8.12) \qquad \operatorname*{res}_{s=1} L_{\mathfrak{a}j}(s) = 8 \, |F|^{-1}$$

by Proposition 6.13. Note that $\Theta_j(1) = 1$ and $\|u_j\| = 1$.

The series (8.9) was introduced independently by Rankin and Selberg (first in the context of classical cusp forms). The Rankin-Selberg L-function inherits a functional equation from that for the Eisenstein series. Precisely, by (8.10) and (6.22) the column vector

$$(8.13) \qquad L_j(s) = [\ldots, L_{\mathfrak{a}j}(s), \ldots]^t$$

satisfies

$$(8.14) \qquad \Theta_j(s)\, L_j(s) = \Phi(s)\, \Theta_j(1-s)\, L_j(1-s)$$

where $\Phi(s)$ is the scattering matrix for the group Γ.

On the vertical lines $\operatorname{Re} s = \sigma > 1$ we get estimates for $L_j(s)$ by (8.7). Hence the functional equation will give us control of the growth of $L_j(s)$ in the critical strip by means of the Phragmen-Lindelöf convexity principle, provided we can control the growth of $\Phi(s)$. The latter is an open problem for general groups. When this problem is resolved in a number of cases, it proves that $(s-1)\, L_{\mathfrak{a}j}(s) \ll |s|^A$ in $\operatorname{Re} s \geq 1 - \varepsilon$. Hence, one obtains by a standard complex integration the following asymptotic formula:

$$(8.15) \qquad \sum_{|n| \leq X} |\nu_{\mathfrak{a}j}(n)|^2 \sim 8\,|F|^{-1}\, X,$$

as $X \to +\infty$. In particular, this says that $|\nu_{\mathfrak{a}j}(n)|$ is about $2\,|F|^{-1/2}$ on average.

In the case of $\Gamma = \Gamma_0(N)$ one can be more precise, because the scattering matrix is computed explicitly (see [He1], [Hu1]). One can prove that $(s-1)\, L_{\mathfrak{a}j}(s)$ is holomorphic in $\operatorname{Re} s \geq 1/2$, and

$$(8.16) \qquad (s-1)\, L_{\mathfrak{a}j}(s) \ll |s\, s_j\, N|^3.$$

Hence by a contour integration together with (8.7) one derives

$$(8.17) \qquad \sum_{|n| \leq X} |\nu_{\mathfrak{a}j}(n)|^2 = 8\,|F|^{-1} X + O(|s_j|\, N\, X^{7/8}),$$

where the implied constant is absolute.

8.3. Bounds for linear forms.

In numerous applications one needs bounds not necessarily for the individual $\nu_{\mathfrak{a}j}(n)$ but for mean values of some kind with respect to n, the spectrum, or both. Sometimes even the group Γ can vary.

In practice one meets linear forms

$$(8.18) \qquad L_{\mathfrak{a}j}(\mathbf{a}) = \sum_{1}^{N} a_n\, \nu_{\mathfrak{a}j}(n)$$

with some complex $\mathbf{a} = (a_n)$. However, one cannot often take advantage of having a special combination; therefore, we might as well consider linear forms in general and seek estimates in terms of the ℓ_2-norm

$$\|\mathbf{a}\|^2 = \sum_1^N |a_n|^2.$$

Immediately, from (8.7) and Cauchy's inequality we derive

$$(8.19) \qquad L_{\mathfrak{a}j}^2(\mathbf{a}) \ll (1 + \lambda_j^{-1}) \left(|s_j| + \frac{N}{c_{\mathfrak{a}}} \right) \|\mathbf{a}\|^2.$$

This estimate is best possible apart from the implied constant, which is absolute here.

Yet, for special linear forms, one should be able to improve upon the individual bound (8.19) by exploiting the variation in sign of $\nu_{\mathfrak{a}j}(n)$. For example, if the a_n are given by additive characters (observe that the Fourier coefficients of a Maass form are determined only up to the twist by a fixed additive character, because the scaling matrix $\sigma_{\mathfrak{a}}$ can be altered by a translation from the right side), we shall prove the following:

Theorem 8.1. *If u_j is a cusp form, then for any real α we have*

$$(8.20) \qquad \sum_{|n| \leq N} e(\alpha n)\, \nu_{\mathfrak{a}j}(n) \ll (\lambda_j N)^{1/2} \log 2N,$$

where the implied constant depends only on the group.

Proof. We have

$$\rho_{\mathfrak{a}j}(n)\, W_{s_j}(iny) = \int_0^1 u_j(\sigma_a z)\, e(-nx)\, dx.$$

Integrating with respect to the measure $y^{-1}\, dy$, we get

$$(8.21) \qquad \pi^{-1/2} \Gamma\Big(\frac{s_j}{2}\Big)\, \Gamma\Big(\frac{1 - s_j}{2}\Big)\, \rho_{\mathfrak{a}j}(n) = \int_0^1 \varphi_{\mathfrak{a}j}(x)\, e(-nx)\, dx,$$

where

$$(8.22) \qquad \varphi_{\mathfrak{a}j}(x) = \int_0^{+\infty} u_j(\sigma_a z)\, y^{-1}\, dy \ll |s_j|^{1/2}$$

with the implied constant depending on the group. For the proof of (8.22), note that $u_j(\sigma_a z)$ is a cusp form for the group $\sigma_{\mathfrak{a}}^{-1} \Gamma \sigma_{\mathfrak{a}}$, so the upper bound (8.3') holds for $u_j(\sigma_a z)$. Summing over n, we get

$$\pi^{-1/2} \Gamma\Big(\frac{s_j}{2}\Big)\, \Gamma\Big(\frac{1 - s_j}{2}\Big) \sum_{|n| \leq N} \rho_{\mathfrak{a}j}(n) = \int_0^1 F_N(x)\, \varphi_{\mathfrak{a}j}(x)\, dx,$$

with the kernel

$$F_N(x) = \sum_{|n| \leq N} e(nx) = \frac{\sin \pi (2N+1)x}{\sin \pi x}.$$

The $\mathfrak{L}^1$-norm of $F_N(x)$ is small, namely

$$\int_0^1 |F_N(x)|\, dx \ll \log 2N,$$

and by applying Stirling's approximation to the gamma factors we infer that

$$\sum_{|n| \leq N} \nu_{\mathfrak{a}j}(n)\, |n|^{-1/2} \ll |s_j|^{1/2} |1 - s_j|^{1/2} \log 2N.$$

Finally, relax $|n|^{-1/2}$ by partial summation and introduce the character $e(\alpha n)$ by changing $\sigma_\mathfrak{a}$ to $\sigma_\mathfrak{a} n(\alpha)$ to complete the proof.

Remarks. The estimate (8.22) is quite crude; probably $\varphi_{\mathfrak{a}j} \ll |s_j|^\varepsilon$. The exponent $1/2$ in (8.20) cannot be reduced for all α in view of Parseval's identity. Nevertheless, for special α and for a properly chosen scaling matrix $\sigma_\mathfrak{a}$ a considerable improvement is possible, if Γ is a congruence group at any rate.

The uniformity in α allows us to obtain interesting consequences. For example, by way of additive characters to modulus q, we can stick to an arithmetic progression, getting the same bound as (8.20), *i.e.*

$$(8.23) \qquad \sum_{\substack{|n| \leq N \\ n \equiv a (\mathrm{mod}\ q)}} \nu_{\mathfrak{a}j}(n) \ll (\lambda_j N)^{1/2} \log 2N$$

for any a and $q \geq 1$, the implied constant depending only on the group.

Lots of other results stem from that simple idea of writing $\rho_{\mathfrak{a}j}(n)$ as the Fourier transform of $u_j(\sigma_\mathfrak{a} z)$. Here is a cute estimate for a bilinear form of additive convolution type.

Theorem. *Let u_j be a cusp form. For any $\mathbf{a} = (a_m)$, $\mathbf{b} = (b_n)$, we have*

$$\sum_{|m| \leq M} \sum_{|n| \leq N} a_m b_n \rho_{\mathfrak{a}j}(m-n) \ll \lambda_j^{1/2} \cosh(\frac{\pi}{2} t_j)\, \|\mathbf{a}\|\, \|\mathbf{b}\|,$$

where the implied constant depends only on the group.

W. Duke pointed out to me that this theorem follows immediately by (8.21), (8.22) and Parseval's identity.

A vast amount of cancellation between terms of the above bilinear form indicates strongly that the Fourier coefficients of a cusp form are in no way near to being an additive character. They rather tend to be multiplicative after an adequate diagonalization by Hecke characters, if Γ is a congruence group. A multiplicative analogue of our theorem in this case is obviously false.

8.4. Spectral mean-value estimates.

Next we establish estimates for the Fourier coefficients of Maass forms on average with respect to the spectrum. Many useful estimates on average with respect to the spectrum for the congruence group $\Gamma_0(q)$ have been established by J.-M. Deshouillers and H. Iwaniec [De-Iw]. Here is a sample:

$$\sum_{|t_j|<T} |L_{\mathfrak{a}j}(\mathbf{a})|^2 \ll (T^2 + q^{-1}N \log 2N) \|\mathbf{a}\|^2$$

for any $N, T \geq 1$ (originally we had N^ε in place of $\log 2N$). In these lectures we prove a result of similar type (see (8.25)), which is somewhat weaker, but it is neat, it holds for any group, and it is useful in general. We also take this opportunity to present more direct arguments which avoid Kloosterman sums.

First, however, let us show what can be inferred from Proposition 7.2, which did not require the spectral theorem in its full force—only Bessel's inequality was used. Integrating (7.10) over a horocycle segment in the $\mathfrak{a}$ cuspidal zone at height y, one gets (apply the Fourier expansions (8.1) and (8.2))

(8.24)
$$\sum_{|t_j|<T} \sum_{n\neq 0} |\rho_{\mathfrak{a}j}(n)\, W_{s_j}(nz)|^2$$
$$+ \sum_{\mathfrak{c}} \frac{1}{4\pi} \int_{-T}^{T} \sum_{n\neq 0} |\varphi_{\mathfrak{a}\mathfrak{c}}(n, s)\, W_s(nz)|^2 \, dt \ll T^2 + yT,$$

for $T \geq 1$ and $z \in \mathbb{H}$, where the implied constant depends on the group.

For fixed n we shall do better by an appeal to the complete spectral decomposition of a particular automorphic kernel (7.17). We begin by estimating the twisted Maass forms which are obtained from the Fourier expansion (8.1) by multiplying its coefficients with a sequence $\mathbf{a} = (a_n)$. Denote these by

$$\mathbf{a} \otimes u_j(\sigma_{\mathfrak{a}}z) = \sum_n a_n\, \rho_{\mathfrak{a}j}(n)\, W_{s_j}(nz)$$

if u_j is a cusp form. Similarly, we twist the Eisenstein series and their residues. For notational simplicity and without loss of generality we are going to consider $\mathfrak{a} = \infty$, $\sigma_{\mathfrak{a}} = 1$ (change the group to $\sigma_{\mathfrak{a}}^{-1}\Gamma\sigma_{\mathfrak{a}}$) and $z = iy$ (change a_n to $a_n e(-nx)$). In this case we also drop the subscript $\mathfrak{a}$ in relevant places and set $A_j(y) = \mathbf{a} \otimes u_j(\sigma_{\mathfrak{a}}z)$. We have

$$A_j(y) = \int_0^1 S(x)\, u_j(z)\, dx,$$

where

$$S(x) = \sum_n a_n\, e(-nx).$$

The Parseval formula asserts (we assume that $\|\mathbf{a}\| < +\infty$)

$$\int_0^1 |S(x)|^2\, dx = \sum_n |a_n|^2 = \|\mathbf{a}\|^2.$$

By the spectral decomposition (7.17), we get

$$\sum_j h(t_j)\, A_j(y)\, \overline{A}_j(y') + \cdots = \int_0^1\int_0^1 S(x)\, \overline{S}(x')\, K(z, z')\, dx\, dx'.$$

Here and hereafter, the three dots stand for the corresponding contribution of the continuous spectrum. Furthermore, this is bounded by

$$\int_0^1\int_0^1 |S(x)|^2 |K(z, z')|\, dx\, dx' = \int_0^1 |S(x)|^2 H(z, y')\, dx,$$

where

$$H(z, y') = \int_0^1 |K(z, z')|\, dx'.$$

Suppose that $k(u)$ is real and non-negative, so that we can drop the absolute value in the automorphic kernel $K(z, z')$. Then $H(z, y')$ is easily recognized as an incomplete Eisenstein series (see the remarks concluding Section 4.2)

$$H(z, y') = \sum_{\gamma \in \Gamma_{\mathfrak{a}}\backslash\Gamma} \psi(\operatorname{Im} \gamma z)$$

(remember that $\mathfrak{a} = \infty$), where $\psi(y) = (y'y)^{1/2}g(\log y'/y)$ and $g(r)$ is the Fourier transform of $h(t)$. The identity motion yields $\psi(y)$. The other motions, by Lemma 2.10, using partial summation contribute all together at most

$$\frac{10}{c_{\mathfrak{a}}} \int_0^{+\infty} y^{-1} |\psi'(y)|\, dy = \frac{10}{c_{\mathfrak{a}}} \int_{-\infty}^{+\infty} e^{r/2} \left| \frac{1}{2}\, g(r) - g'(r) \right|\, dr.$$

To simplify, assume that $g(r)$ is positive and decreasing on $\mathbb{R}^+$ (this hypothesis implies our former condition that $k(u)$ is non-negative). Then the last integral is bounded by

$$\int_0^{+\infty} (g(r) - 2\,g'(r)) \cosh \frac{r}{2}\, dr = g(0) + h\left(\frac{i}{2}\right).$$

Therefore,

$$H(z, y') \leq (y'y)^{1/2} g\left(\log \frac{y'}{y}\right) + \frac{10}{c_{\mathfrak{a}}}\left(g(0) + h\left(\frac{i}{2}\right)\right).$$

Taking $y' = y$, we arrive at the following inequality:

$$\sum_j h(t_j)\, |\mathbf{a} \otimes u_j(\sigma_{\mathfrak{a}} z)|^2 + \cdots \leq \left(\left(y + \frac{10}{c_{\mathfrak{a}}}\right) g(0) + \frac{10}{c_{\mathfrak{a}}} h\left(\frac{i}{2}\right)\right) \|\mathbf{a}\|^2.$$

For the Fourier pair $h(t) = \exp(-t^2/4T^2)$, $g(r) = \pi^{-1/2} T \exp(-r^2 T^2)$ this yields

Theorem 8.2. *For $T \geq 1$ and any complex $\mathbf{a} = (a_n)$ with $\|\mathbf{a}\| < +\infty$*

(8.25)
$$\sum_{|t_j| < T} |\mathbf{a} \otimes u_j(\sigma_{\mathfrak{a}} z)|^2 + \sum_{\mathfrak{c}} \frac{1}{4\pi} \int_{-T}^{T} |\mathbf{a} \otimes E_{\mathfrak{c}}(\sigma_{\mathfrak{a}} z, \tfrac{1}{2} + it)|^2\, dt$$
$$< T(y + 22\,c_{\mathfrak{a}}^{-1})\, \|\mathbf{a}\|^2.$$

Now take $\mathbf{a} = (a_n)$ in which all but one entry vanishes, getting

(8.26)
$$\sum_{|t_j| < T} |\rho_{\mathfrak{a}j}(n)\, W_{s_j}(nz)|^2$$
$$+ \sum_{\mathfrak{c}} \frac{1}{4\pi} \int_{-T}^{T} |\varphi_{\mathfrak{a}\mathfrak{c}}(n, s)\, W_s(nz)|^2\, dt < T(y + 22\,c_{\mathfrak{a}}^{-1})$$

for $T \geq 1$ and $z \in \mathbb{H}$. Compare this with (8.24).

Next clear (8.26) of the Whittaker functions by integrating over the dyadic interval $Y < y < 2Y$ with $Y = cT/|n|$ (as in the proof of Theorem 3.2 the integration is necessary, because one cannot find a universal value of y for which all $|W_{s_j}(iny)|$ are simultaneously well bounded below). One gets the following estimate:

(8.27)
$$\sum_{|t_j| < T} |\nu_{\mathfrak{a}j}(n)|^2 + \sum_{\mathfrak{c}} \frac{1}{4\pi} \int_{-T}^{T} |\eta_{\mathfrak{a}\mathfrak{c}}(n, t)|^2\, dt \ll T^2 + c_{\mathfrak{a}}^{-1}\, |n|\, T$$

for any $T \geq 1$ and $n \neq 0$, where the implied constant is absolute. A more precise asymptotic formula (9.13) will be derived from the spectral formula with Kloosterman sums (9.12); still, (8.27) has the advantage of being uniform and sharp with respect to the group Γ.

8.5. The case of congruence groups.

The general bounds so far established are not bad if one considers the uniformity in present parameters. However, for congruence groups some estimates can be improved. What makes this possible is the existence of a special basis in $\mathfrak{L}(\Gamma\backslash\mathbb{H})$ which diagonalizes the Hecke operators. For the classical automorphic forms this is the core of the Atkin-Lehner theory of newforms [At-Le]. The case of Maass forms is identical except for verbal differences. We do not wish to develop this theory from scratch here, but rather only transcribe briefly the main concepts and results for the group $\Gamma_0(N)$.

For $n \geq 1$, define the set

$$(8.28) \qquad \Gamma_n = \left\{ \begin{pmatrix} a & b \\ c & d \end{pmatrix} : a,b,c,d \in \mathbb{Z},\ ad - bc = n \right\}.$$

In particular, Γ_1 is the modular group. Naturally Γ_1 acts on Γ_n. The Hecke operator $T_n : \mathcal{A}(\Gamma_1\backslash\mathbb{H}) \longrightarrow \mathcal{A}(\Gamma_1\backslash\mathbb{H})$ is defined by

$$(8.29) \qquad (T_n f)(z) = \frac{1}{\sqrt{n}} \sum_{\tau \in \Gamma_1\backslash\Gamma_n} f(\tau z).$$

Picking up specific representatives of $\Gamma_1\backslash\Gamma_n$, we can also write

$$(8.30) \qquad (T_n f)(z) = \frac{1}{\sqrt{n}} \sum_{ad=n} \sum_{b(\text{mod } d)} f\left(\frac{az+b}{d}\right).$$

Clearly, this sum is finite, the number of terms being

$$[\Gamma_n : \Gamma_1] = \sum_{d|n} d = \sigma(n),$$

Therefore, T_n is bounded on $\mathfrak{L}(\Gamma_1\backslash\mathbb{H})$ by $\sigma(n)\, n^{-1/2}$. One can prove the following multiplication rule:

$$(8.31) \qquad T_m T_n = \sum_{d|(m,n)} T_{mnd^{-2}},$$

so that in particular T_m and T_n commute. Moreover, the Hecke operators commute with the Laplace operator, because they are defined by the group operations.

Consider the Hecke congruence group $\Gamma_0(N)$ of level $N \geq 1$. Since $\mathcal{A}(\Gamma_0(N)\backslash\mathbb{H}) \subset \mathcal{A}(\Gamma_1\backslash\mathbb{H})$, every operator T_n acts on $\mathcal{A}(\Gamma_0(N)\backslash\mathbb{H})$. Nevertheless, only those T_n with $(n, N) = 1$ are interesting. First of all, T_n is self-adjoint in $\mathfrak{L}(\Gamma_0(N)\backslash\mathbb{H})$, *i.e.*

$$\langle T_n f, g \rangle = \langle f, T_n g \rangle, \qquad \text{if } (n, N) = 1$$

(the other operators are not even normal). Therefore, in the space of cusp forms $\mathcal{C}(\Gamma_0(N)\backslash\mathbb{H})$ an orthogonal basis $\{u_j(z)\}$ can be chosen which consists of simultaneous eigenfunctions for all T_n, *i.e.*

$$(8.32) \qquad T_n u_j(z) = \lambda_j(n)\, u_j(z), \qquad \text{if } (n, N) = 1.$$

The Eisenstein series $E_\infty(z, 1/2 + it)$ is shown to be an eigenfunction of all the Hecke operators T_n, $(n, N) = 1$, with eigenvalue

$$(8.33) \qquad \eta_t(n) = \sum_{ad=n} \left(\frac{a}{d}\right)^{it}.$$

It is conjectured (Ramanujan-Petersson) that

$$(8.34) \qquad |\lambda_j(n)| \le \tau(n), \qquad (n, N) = 1.$$

By (8.33) the conjecture is obviously true in the space of continuous spectrum. In the cuspidal space the following bound was derived by using estimates for hyper-Kloosterman sums (among other things of arithmetical nature):

$$\lambda_j(n) \le \tau(n)\, n^{5/28}.$$

This is due to D. Bump, W. Duke, J. Hoffstein, H. Iwaniec [Bu-Du-Ho-Iw]. (At the time of preparing this book for the AMS edition the best result is due to H. Kim and P. Sarnak [Ki-Sa], the exponent 5/28 being replaced by 7/64.) For the constant eigenfunction the Hecke eigenvalue is much larger; we have exactly

$$(8.35) \qquad \lambda_0(n) = \sigma(n)\, n^{-1/2}.$$

By virtue of (8.32) the Hecke operator T_n acts on the Fourier coefficients of $u_j(z)$ in the cusp $\mathfrak{a} = \infty$ simply by a constant factor:

$$(8.36) \qquad \nu_j(n) = \nu_j(1)\, \lambda_j(n), \qquad \text{if } (n, N) = 1.$$

In fact, it follows from (8.30)-(8.32) that

$$(8.37) \qquad \nu_j(m)\, \lambda_j(n) = \sum_{d|(m,n)} \nu_j\left(\frac{mn}{d^2}\right), \qquad \text{if } (n, N) = 1.$$

From now on we drop the subscript j for notational simplicity. The relation (8.36) says that the Fourier coefficients $\nu(n)$ are proportional to the Hecke eigenvalues $\lambda(n)$ provided $\nu(1) \ne 0$, but unfortunately $\nu(1)$ may vanish for some forms. For example, take a cusp form $v(z)$ on an overgroup $\Gamma_0(M) \supset$

$\Gamma_0(N)$ with $M|N$. Then $u(z) = v(Dz)$, where $DM|N$, is a cusp form on $\Gamma_0(N)$ all of whose coefficients $\nu(n)$ vanish, save for $n \equiv 0 \pmod{D}$. If $M < N$, such a $v(Dz)$ is seen as an oldform. Atkin and Lehner have shown how to split the space of cusp forms into newforms. Let us write

$$\mathcal{C}(\Gamma_0(N)\backslash\mathbb{H}) = \mathcal{C}_{\mathrm{old}}(\Gamma_0(N)\backslash\mathbb{H}) \oplus \mathcal{C}_{\mathrm{new}}(\Gamma_0(N)\backslash\mathbb{H}).$$

Here $\mathcal{C}_{\mathrm{old}}(\Gamma_0(N)\backslash\mathbb{H})$ is the linear subspace of $\mathcal{C}(\Gamma\backslash\mathbb{H})$ spanned by forms of type $v(Dz)$, where $v(z)$ is a Maass cusp form on $\Gamma_0(M)$ with $DM|N$, $M < N$, and by definition $\mathcal{C}_{\mathrm{new}}(\Gamma_0(N)\backslash\mathbb{H})$ is the orthogonal complement. Clearly T_n with $(n, N) = 1$ maps $\mathcal{C}_{\mathrm{old}}(\Gamma_0(N)\backslash\mathbb{H})$ into itself, because it commutes with the operator $f(z) \mapsto f(Dz)$ for every $D|N$. Consequently, T_n maps $\mathcal{C}_{\mathrm{new}}(\Gamma_0(N)\backslash\mathbb{H})$ into itself, because T_n is hermitian. Therefore, there exists a basis in $\mathcal{C}_{\mathrm{new}}(\Gamma_0(N)\backslash\mathbb{H})$ of Maass cusp forms which are common eigenfunctions of all T_n with $(n, N) = 1$ (one can work it out in each spectral eigenspace separately). These cusp forms are called *newforms* of level N. The newforms are the GL_2 analogues of the primitive Dirichlet characters $\chi(\mathrm{mod}\ N)$.

Let us return to the space $\mathcal{C}_{\mathrm{old}}(\Gamma_0(N)\backslash\mathbb{H})$. A function $u(z)$ in this space is called an *oldform* if $u(z) = v(Dz)$, where $v(z)$ is a newform on some overgroup $\Gamma_0(M)$ with $DM|N$, $M < N$. In this case we say that $u(z)$ is an *oldform of level M and divisor D*. It turns out, but it is not automatically a fact, that the space $\mathcal{C}_{\mathrm{old}}(\Gamma_0(N)\backslash\mathbb{H})$ is spanned by oldforms.

Another pleasant fact is that a newform of level N, besides being a common eigenfunction of all T_n with $(n, N) = 1$, is automatically an eigenfunction of all the operators U_p, $p|N$, defined by

$$(8.38) \qquad (U_p f)(z) = \frac{1}{\sqrt{p}} \sum_{b(\mathrm{mod}\ p)} f\left(\frac{z+b}{p}\right).$$

This fact follows from the multiplicity-one property (see [At-La]).

The main profit from splitting the space of cusp forms into newforms is the multiplicativity of the Fourier coefficients. Precisely, if $u(z)$ is a newform on $\Gamma_0(N)$, then its first Fourier coefficient $\nu(1)$ does not vanish (it is customary to normalize $u(z)$ by setting $\nu(1) = 1$, which we reject in favor of the $\mathcal{L}^2$-normalization to avoid confusion), and $\lambda(n) = \nu(n)/\nu(1)$ satisfy the following rules of multiplication:

$$(8.39) \qquad
\begin{aligned}
\lambda(m)\,\lambda(n) &= \sum_{d|(m,n)} \lambda\left(\frac{mn}{d^2}\right) & \text{if } (n, N) = 1, \\
\lambda(m)\,\lambda(p) &= \lambda(mp) & \text{if } p|N.
\end{aligned}$$

Observe that for all $m, n \geq 1$ one has

$$(8.40) \qquad |\lambda(m)\,\lambda(n)| \leq \sum_{d|(m,n)} \left| \lambda\left(\frac{mn}{d^2}\right) \right|.$$

Trivially, $|\nu(n)| = |\nu(1)\,\lambda(n)| \leq |\nu(1)|\,\lambda_0(n)$; hence by (8.17) we get a crude lower bound

$$(8.41) \qquad |\nu(1)|^2 \gg \lambda^{-8}\, N^{-33}.$$

Now we are ready to use the power of multiplication to establish the following:

Theorem 8.3. *Let $u(z)$ be a newform on $\Gamma_0(N)$ with eigenvalue λ and Fourier coefficients $\nu(n) = \nu(1)\,\lambda(n)$. Then we have*

$$(8.42) \qquad x^{1/2}\,(\log 2Nx)^{-1} \ll \sum_{0 < n \leq x} |\lambda(n)|^2 \ll x\,(\lambda\,N)^\varepsilon$$

for all $x \geq 1$ and any $\varepsilon > 0$. Moreover, we have

$$(8.43) \qquad N^{-1}(\lambda\,N)^{-\varepsilon} \ll |\nu(1)|^2 \ll \lambda^{1/4}N^{-1/2}\log 2N ,$$

the implied constant depending only on ε.

Remarks. The upper bound of (8.42) follows easily from the Ramanujan-Petersson conjecture (8.4), but this important conjecture is beyond the reach of current methods. By (8.7) we have

$$(8.44) \qquad \sum_{0 < n \leq x} |\nu(n)|^2 \ll x\,N^{-1} + \lambda^{1/2},$$

where the implied constant is absolute (assuming $\|u\| = 1$). Therefore, the novelty of (8.42) rests in the short range of $x < \lambda^{1/2}N$, which is crucial for applications.

Proof. The lower bound of (8.42) is easy to prove; just observe that if $p \nmid N$, then either $|\lambda(p)| \geq 1/2$ or $|\lambda(p^2)| \geq 1/2$, since $\lambda(p)^2 = \lambda(p^2) + 1$ by (8.39). For the proof of the upper bound of (8.42) we consider the Rankin-Selberg L-function

$$L(s) = \sum_1^\infty |\lambda(n)|^2 n^{-s}$$

for $1 < s < 2$. First, by (8.44) and (8.41), we derive a crude estimate

$$L(s) \ll |\nu(1)|^{-2}\lambda^{1/2}(s-1)^{-1} \ll (\lambda N)^{33}(s-1)^{-1},$$

where the implied constant is absolute. Then by (8.40) and Cauchy's inequality we infer that

$$L^2(s) \leq \sum_{m,n} \left(\sum_{d|(m,n)} \left| \lambda\left(\frac{mn}{d^2}\right) \right| \right)^2 (mn)^{-s}$$

$$\leq \sum_{m,n} \tau((m,n)) \sum_{d|(m,n)} \left| \lambda\left(\frac{mn}{d^2}\right) \right|^2 (mn)^{-s}$$

$$= \sum_{\ell} |\lambda(\ell)|^2 \ell^{-s} \sum_{\substack{d\,m\,n \\ mn=\ell}} \tau((m,n)\,d)\, d^{-2s}$$

$$\leq \zeta^2(2s) \sum_{\ell} |\tau(\ell)\,\lambda(\ell)|^2 \ell^{-s} \ll L(s-\varepsilon),$$

because $\tau(\ell)^2 \ll \ell^{\varepsilon}$ for any $0 < \varepsilon < s - 1$, the implied constant depending on ε alone. Iterating the obtained inequality, we get

$$L^{2^k}(s) \ll L(s - \varepsilon k) \ll (\lambda N)^{33}$$

for any $k \geq 1$ and $0 < \varepsilon k < s - 1$, the implied constant depending only on ε, s and k. Taking the root of degree 2^k with k large, we get the bound

(8.45) $L(s) \ll (\lambda N)^{\varepsilon}$

for any $\varepsilon > 0$ and $s > 1$, the implied constant depending on ε and s. This bound implies (8.42) with the extra factor x^{ε}, which can be removed by means of (8.44).

The lower bound for $|\nu(1)|^2$ in (8.43) follows immediately by comparing the upper bound of (8.42) with (8.17) for $x = (\lambda N)^{33}$. The upper bound for $|\nu(1)|^2$ is obtained by combining (8.44) with the lower bound of (8.42) for $x = \lambda^{1/2} N$.

Presumably, the lower bound (8.42) should be $x\,(\lambda N)^{-\varepsilon}$ uniformly in $x \geq 1$, but this resists a proof. Such a bound is desired for numerous applications (see for example Section 13.4). It would give us, among other things, the following estimate:

(8.46) $|\nu(1)|^2 \ll N^{-1}(\lambda N)^{\varepsilon}.$

It turns out that $N^{-1} |\nu(1)|^{-2}$ is (up to a constant factor) the symmetric square L-function associated with the cusp form $u(z)$ at the point $s = 1$. Hence the upper bound (8.46) translates into a lower bound for the symmetric square L-function. Very recently J. Hoffstein and P. Lockhart [Ho-Lo] have established such a bound (in fact a slightly better one) unconditionally, using quite advanced results from the theory of automorphic L-functions of high rank in the style of E. Landau [L] (see also the appendix to [Ho-Lo] by D. Goldfeld, J. Hoffstein and D. Lieman). The estimates (8.43) and (8.46) combined determine the true size of $\nu(1)$.

Chapter 9

Spectral Theory of Kloosterman Sums

9.1. Introduction.

Kloosterman sums were invented to refine the circle method of Hardy and Ramanujan [Ha-Ra]. Originally, Kloosterman [Kl1] applied his refinement to counting representations by a quadratic form in four variables. Shortly afterwards, Kloosterman [Kl2] and Rademacher [Rad] used the idea to estimate the Fourier coefficients of classical modular forms. These coefficients were later expressed effectively as sums of Kloosterman sums without appealing to the circle method by H. Petterson [Pe1], R. Rankin [Ran] and A. Selberg [Se3] independently. Therefore, a connection between Kloosterman sums and modular forms was established right away. Next, algebraic geometry became associated with modular forms via Weil's estimate for Kloosterman sums (a special case of the Riemann hypothesis for curves). Consequently, the results became deeper, but still not the best possible. Only recently has a complete picture emerged from the spectral theory of automorphic forms. Its essence is captured in the analytic continuation of the series

$$(9.1) \qquad L_s(m,n) = \sum_{c>0} c^{-2s} \mathcal{S}_{\mathfrak{ab}}(m,n;c),$$

due to A. Selberg [Se1]. There is an elegant treatment of this series by D. Goldfeld and P. Sarnak [Go-Sa]. In these lectures we catch a glimpse of that profound theory. Instead of (9.1) we shall examine the series (5.16), which occurs on our way more naturally. Of course both are related via the power series expansion for the Bessel functions $J_{2s-1}(x)$ and $I_{2s-1}(x)$; precisely, we have

$$(9.2) \qquad Z_s(m,n) = \pi (4\pi^2 |mn|)^{s-1} \sum_{k=0}^{\infty} \frac{(4\pi^2 mn)^k}{k!\,\Gamma(k+2s)} \, L_{s+k}(m,n) \,.$$

Recall that the series $L_s(m,n)$ and $Z_s(m,n)$ converge absolutely in $\mathrm{Re}\,s > 1$.

9.2. Analytic continuation of $Z_s(m,n)$.

We shall appeal to the properties of the automorphic Green function $G_s(z/z')$ already established (originally Selberg employed Poincaré series). The idea is straightforward: on one hand the zeta-function $Z_s(m,n)$ turns up in the Fourier expansion (5.15); on the other hand we have the spectral decomposition (7.18) for the difference $G_s(z/z') - G_a(z/z')$ with $a > 1$, $\mathrm{Re}\,s > 1$ in terms of Maass forms. These forms have Fourier expansions too (see (8.1) and (8.2)). Comparing the (m,n)-th Fourier coefficients for $mn \neq 0$ of both expansions, we are led to the following identity:

$$\delta_{\mathfrak{ab}}\,\delta_{mn}\,(4\pi|n|)^{-1} W_s(iny')\,V_s(iny) + Z_s(m,n)\,W_s(imy')\,W_s(iny)$$
$$- \text{the same expresion for } s = a$$

$$(9.3)$$
$$= \sum_j \chi_{sa}(s_j)\,\bar{\rho}_{\mathfrak{a}j}(m)\,\rho_{\mathfrak{b}j}(n)\,W_{s_j}(imy')\,W_{s_j}(iny)$$
$$+ \text{the countinuous spectrum integrals,}$$

where we recall that

$$\chi_{sa}(v) = (s-v)^{-1}(1-s-v)^{-1} - (a-v)^{-1}(1-a-v)^{-1}.$$

To be precise this step requires the absolute convergence of the relevant hybrid Fourier/spectral series. The upper bound (8.24) is just adequate to verify the convergence in question, so (9.3) is established thoroughly. Another point is that the Fourier expansion (5.15) is valid only for (z,z') in the set $D_{\mathfrak{ab}}$; therefore (9.3) requires $y' > y$, $y'y > c(\mathfrak{a},\mathfrak{b})^{-2}$, and these conditions are indispensable (we can allow weak inequalities by continuity).

The relation (9.3) will simplify considerably if we let $a \to +\infty$, since the corresponding terms vanish. To see that we can take the limit on the left hand side, we apply the trivial bound $O(c^2)$ for the Kloosterman sums $S_{\mathfrak{ab}}(m,n;c)$, and for the Bessel functions we apply the asymptotics

$$J_\nu(x) \sim \left(\frac{x}{2}\right)^\nu \Gamma(\nu+1)^{-1} \qquad \text{and} \qquad K_\nu(x) \sim \left(\frac{2}{x}\right)^\nu \Gamma(\nu)$$

as $\nu \to \infty$ uniformly in $0 < x \ll 1$ (see the power series expansions). Hence, we infer that

$$W_s(iny')\,V_s(iny) \ll \frac{\Gamma(s-1/2)}{\Gamma(s+1/2)} = \left(s - \frac{1}{2}\right)^{-1},$$

$$W_s(imy')\,W_s(iny)\,J_{2s-1}\!\left(\frac{4\pi\sqrt{|mn|}}{c}\right)$$
$$\ll (c^2 y'y)^{1/2-s}\,\frac{\Gamma(s-1/2)^2}{\Gamma(2s)} \ll c^{-3}\,s^{-3/2},$$

if $y'y \geq c^{-2}$ and $s \geq 2$. These estimates also hold true with J_{2s-1} replaced by I_{2s-1}. Therefore, the contribution to the left side of (9.3) of terms for $s = a$ is bounded by $O(a^{-1})$. On the right side we split the series of spectral terms in accordance with $\chi_{sa}(v)$ and show that the resulting series for $s = a$ vanish as $a \to +\infty$. To be correct, one ought to verify the uniform convergence. For this purpose the bound (8.26) is more than sufficient, while (8.24) barely misses.

This said, in (9.3) we drop all terms with $s = a$. Furthermore, for simplicity we change $2\pi|n|y \to y$, $2\pi|m|y' \to y'$, and to obtain symmetry we set

$$(9.4) \qquad 2\,D_\nu(y,y') = \begin{cases} I_\nu(y)\,K_\nu(y'), & \text{if } y' \geq y, \\ I_\nu(y')\,K_\nu(y), & \text{if } y' \leq y. \end{cases}$$

We obtain

Proposition 9.1. *Suppose $y'y \geq 4\pi^2 c(\mathfrak{a},\mathfrak{b})^{-2}|mn|$. For any s with $\mathrm{Re}\,s > 1$ we have*

$$\frac{1}{2|n|}\,\delta_{\mathfrak{a}\mathfrak{b}}\,\delta_{mn}\,D_{s-1/2}(y',y) + Z_s(m,n)\,K_{s-1/2}(y')\,K_{s-1/2}(y)$$

$$(9.5) \qquad = \sum_j (s-s_j)^{-1}(1-s-s_j)^{-1}\bar\rho_{\mathfrak{a}j}(m)\,\rho_{\mathfrak{b}j}(n)\,K_{it_j}(y')\,K_{it_j}(y)$$

$$+ \sum_{\mathfrak{c}} \frac{1}{4\pi i}\int_{(1/2)} (s-v)^{-1}(1-s-v)^{-1}\bar\varphi_{\mathfrak{a}\mathfrak{c}}(m,v)\,\varphi_{\mathfrak{b}\mathfrak{c}}(n,v)$$
$$\cdot K_{v-1/2}(y')\,K_{v-1/2}(y)\,dv.$$

The sum over the discrete spectrum in (9.5) extends to all of $\mathbb{C}$ by analytic continuation, and gives a function invariant under the change $s \mapsto 1 - s$. It has simple poles at $s = s_j$ and at $s = 1 - s_j$ with the same residue

$$-(2s_j - 1)^{-1}\sum_{s_k = s_j} \bar\rho_{\mathfrak{a}k}(m)\,\rho_{\mathfrak{b}k}(n)\,K_{it_j}(y')\,K_{it_j}(y),$$

provided $s_j \neq 1/2$. If $s_j = 1/2$ (so $\lambda_j = 1/4$ belongs to the discrete spectrum, which is necessarily cuspidal), then it contributes

$$-(s - \frac{1}{2})^{-2} \sum_{s_k = 1/2} \bar{\rho}_{ak}(m)\, \rho_{bk}(n)\, K_0(y')\, K_0(y)\,.$$

This has a double pole at $s = 1/2$ with residue zero.

The continuous spectrum integrals in (9.5) give functions holomorphic in $\mathrm{Re}\, s > 1/2$. To extend these to $\mathrm{Re}\, s \leq 1/2$ we repeat the arguments given for the Green function in Section 7.4. Accordingly we shall move the integration from $\mathrm{Re}\, v = 1/2$ to $\mathrm{Re}\, v = \alpha$ with α larger but close enough to $1/2$ so that all $\varphi_{ac}(m, v)$, $\varphi_{bc}(n, v)$ are holomorphic in the strip $1/2 < \mathrm{Re}\, v < \alpha$. Assuming for a moment that s is in this strip, we pass the pole at $v = s$ with residue

$$-(2s - 1)^{-1} \varphi_{ac}(m, 1 - s)\, \varphi_{bc}(n, s)\, K_{s-1/2}(y')\, K_{s-1/2}(y)\,.$$

The resulting integral over the vertical line $\mathrm{Re}\, v = \alpha$ gives a holomorphic function in the strip $1 - \alpha < \mathrm{Re}\, s < \alpha$ which is also invariant under the change $s \mapsto 1 - s$. Also invariant except for sign change is the above residue, since we have

$$(9.6) \qquad \sum_{c} \varphi_{ac}(m, 1 - s)\, \varphi_{bc}(n, s) = \sum_{c} \varphi_{ac}(m, s)\, \varphi_{bc}(n, 1 - s)$$

by (7.20). Finally, notice that

$$D_\nu(y', y) - D_{-\nu}(y', y) = -\frac{1}{\pi}\, \sin(\pi\nu)\, K_\nu(y')\, K_\nu(y),$$

since

$$(9.7) \qquad K_\nu(y) = \frac{\pi}{2 \sin \pi\nu} \big(I_{-\nu}(y) - I_\nu(y)\big)\,.$$

Having made the above transformations, we take the difference of (9.5) at s and $1 - s$. We find that in every term left the Bessel functions $K_{it}(y')$, $K_{it}(y)$ match and wash away, leaving us with the clean

Theorem 9.2. *The series $Z_s(m, n)$ has an analytic continuation in s to all $\mathbb{C}$, and it satisfies the functional equation*

$$(9.8) \qquad \begin{aligned} Z_s(m, n) - Z_{1-s}(m, n) &= \frac{1}{2\pi|n|}\, \delta_{ab}\, \delta_{mn}\, \sin \pi\big(s - \frac{1}{2}\big) \\ &\quad - \frac{1}{2s - 1} \sum_{c} \varphi_{ac}(m, 1 - s)\, \varphi_{bc}(n, s)\,. \end{aligned}$$

$Z_s(m,n)$ *has simple poles at* $s = s_j$ *and* $s = 1 - s_j$ *with residue*

$$-\frac{1}{2s_j - 1} \sum_{s_k = s_j} \bar{\rho}_{\mathfrak{a}k}(m)\, \rho_{\mathfrak{b}k}(n)$$

provided $s_j \neq 1/2$. *At* $s = 1/2$ *it has a pole with Taylor expansion*

$$Z_s(m,n) = -\frac{1}{(s - 1/2)^2} \sum_{s_k = 1/2} \bar{\rho}_{\mathfrak{a}k}(m)\, \rho_{\mathfrak{b}k}(n)$$
$$-\frac{1}{4(s - 1/2)} \sum_{\mathfrak{c}} \bar{\varphi}_{\mathfrak{a}\mathfrak{c}}(m, 1/2)\, \varphi_{\mathfrak{b}\mathfrak{c}}(n, 1/2) + \cdots,$$

where the first sum is void if $s_k = 1/2$ *does not exist.*

Remarks. Since the poles of Z_s and Z_{1-s} at $s \neq 1/2$ cancel out, it follows that $(s - 1/2)(Z_s - Z_{1-s})$ is entire, and so is the sum (9.6).

9.3. Bruggeman-Kuznetsov formula.

For practical purposes we are going to capture the analytic properties of the Kloosterman sums zeta-function $Z_s(m,n)$ in a kind of Poisson summation formula with test functions on both sides as flexible as possible. This is routine for the Dirichlet series in analytic number theory which satisfy standard functional equations. Our case of $Z_s(m,n)$ is not much different, except that the formula in question will not be self-dual.

Let $h(t)$ be a test function which satisfies the conditions (1.63). Put $f(s) = 4\pi(s - 1/2)(\sin \pi s)^{-1} h(i(s - 1/2))$. Thus $f(s)$ is holomorphic in the strip $\varepsilon \leq \operatorname{Re} s \leq 1 - \varepsilon$ and bounded by (recall that $s = 1/2 + it$)

$$f(s) \ll (|t| + 1)^{-1-\varepsilon} |\cosh \pi t|^{-1}.$$

Also $f(s) = -f(1 - s)$. Multiply (9.8) through by $f(s)$ and integrate over the vertical line $\operatorname{Re} s = 1 - \varepsilon$, getting

$$\frac{1}{2\pi i} \int\limits_{(1-\varepsilon)} Z_s(m,n)\, f(s)\, ds + \frac{1}{2\pi i} \int\limits_{(\varepsilon)} Z_s(m,n)\, f(s)\, ds$$
$$= \frac{1}{2\pi |n|}\, \delta_{\mathfrak{a}\mathfrak{b}}\, \delta_{mn}\, \frac{1}{2\pi i} \int\limits_{(1-\varepsilon)} \sin \pi \left(s - \frac{1}{2}\right) f(s)\, ds$$
$$- \frac{1}{2\pi i} \int\limits_{(1-\varepsilon)} \sum_{\mathfrak{c}} \varphi_{\mathfrak{a}\mathfrak{c}}(m, 1 - s)\, \varphi_{\mathfrak{b}\mathfrak{c}}(n, s)\, \frac{f(s)}{2s - 1}\, ds.$$

On the right side the diagonal part becomes $-|n|^{-1}\delta_{\mathfrak{ab}}\,\delta_{mn}\,h_0$, where

(9.9) $$h_0 = \frac{1}{\pi}\int_{-\infty}^{+\infty} t\,\tanh(\pi t)\,h(t)\,dt\,,$$

and the continuous spectrum integrals yield

$$\int_{-\infty}^{+\infty}\sum_{\mathfrak{c}}\varphi_{\mathfrak{ac}}\!\left(m,\frac{1}{2}-it\right)\varphi_{\mathfrak{bc}}\!\left(n,\frac{1}{2}+it\right)\frac{h(t)}{\cosh\pi t}\,dt$$

after moving to the critical line. On the left side the second integral is brought to the first one by moving to the line $\operatorname{Re} s = 1-\varepsilon$. By Cauchy's theorem, the poles at $s = s_j$ and $s = 1 - s_j$ contribute

$$-4\pi\sum_{j}\bar{\rho}_{\mathfrak{a}j}(m)\,\rho_{\mathfrak{b}j}(n)\,\frac{h(t_j)}{\cosh\pi t_j}\,,$$

where of each two points s_j, $1 - s_j$ only one counts (check separately the residue at $s_j = 1/2$). When on the line $\operatorname{Re} s = 1-\varepsilon$ we expand $Z_s(m,n)$ into series of Kloosterman sums (see (5.16)) and integrate termwise, getting

$$\frac{2}{2\pi i}\int_{(1-\varepsilon)} Z_s(m,n)\,f(s)\,ds$$

$$= |mn|^{-1/2}\sum_{c}c^{-1}\mathcal{S}_{\mathfrak{ab}}(m,n;c)\,h^{\pm}\!\left(\frac{4\pi\sqrt{|mn|}}{c}\right),$$

where

$$h^{+}(x) = \frac{1}{2\pi i}\int_{(1-\varepsilon)} J_{2s-1}(x)\,f(s)\,ds$$

if $mn > 0$, and $h^{-}(x)$ is the same integral with I_{2s-1} in place of J_{2s-1} if $mn < 0$. Moving to the critical line, we get

(9.10) $$h^{+}(x) = 2i\int_{-\infty}^{+\infty} J_{2it}(x)\,\frac{h(t)\,t}{\cosh\pi t}\,dt$$

and (use (9.7))

(9.11) $$h^{-}(x) = \frac{4}{\pi}\int_{-\infty}^{+\infty} K_{2it}(x)\,h(t)\,\sinh(\pi t)\,t\,dt\,.$$

Remarks. The above presentation requires the absolute convergence of the series for $Z_s(m,n)$ on $\operatorname{Re} s = 1-\varepsilon$ which hypothesis is often satisfied.

Nevertheless, this hypothesis can be avoided by moving to the line $\operatorname{Re} s = 1+\varepsilon$, where the series does converge absolutely. It will produce an additional term $4\,h(i/2)\,Z_1(m,n)$ from the pole of $f(s)$ at $s = 1$. The same term will reappear from evaluation of $h^\pm(x)$ after moving back to $\operatorname{Re} s = 1 - \varepsilon$ (and further down to $\operatorname{Re} s = 1/2$). Therefore this term contributes nothing. At this point one needs the convergence of $Z_1(m,n)$ (not necessarily absolute), which can be established by a routine estimate using (9.5) and (8.26) (employ the holomorphy of $Z_s(m,n)$ at $s = 1$).

From the above parts we assemble the following formula.

Theorem 9.3. *Let* $\mathfrak{a}$, $\mathfrak{b}$ *be cusps of* Γ, $mn \neq 0$, *and* $\nu_{\mathfrak{a}j}(m)$, $\nu_{\mathfrak{b}j}(n)$, $\eta_{\mathfrak{a}\mathfrak{c}}(m,t)$, $\eta_{\mathfrak{b}\mathfrak{c}}(n,t)$ *the scaled Fourier coefficients of a complete orthonormal system of Maass forms and the eigenpacket of Eisenstein series in* $\mathcal{L}(\Gamma\backslash\mathbb{H})$ *(see (8.5), (8.6)). Then for any* $h(t)$ *satisfying (1.63) we have*

$$
\begin{aligned}
\text{(9.12)} \quad &\sum_j h(t_j)\,\bar\nu_{\mathfrak{a}j}(m)\,\nu_{\mathfrak{b}j}(n) \\
&+ \sum_\mathfrak{c} \frac{1}{4\pi} \int_{-\infty}^{+\infty} h(t)\,\bar\eta_{\mathfrak{a}\mathfrak{c}}(m,t)\,\eta_{\mathfrak{b}\mathfrak{c}}(n,t)\,dt \\
&= \delta_{\mathfrak{a}\mathfrak{b}}\,\delta_{mn}\,h_0 + \sum_c c^{-1} S_{\mathfrak{a}\mathfrak{b}}(m,n;c)\,h^\pm\!\left(\frac{4\pi\sqrt{|mn|}}{c}\right),
\end{aligned}
$$

where $\pm$ *is the sign of* mn *and* h_0, h^+, h^- *are the integral transforms of* h *given by (9.9)-(9.11).*

Remarks. One could derive (9.12) by integrating (9.3) instead of (9.8). This formula was established first for the modular group by N. V. Kuznetsov, and in a slightly less refined form by R. W. Bruggeman [Br1]. They used the Selberg-Poincaré series $E_{\mathfrak{a}m}(z,s)$ rather than the Green function. N.V. Proskurin [Pr] considered the general case of finite volume groups and forms of arbitrary weight. Motivated by numerous applications, J.-M. Deshouillers and H. Iwaniec [De-Iw] worked out the formula with mixed cusps for the group $\Gamma_0(q)$. J. Cogdell and I. Piatetski-Shapiro [Co-Pi] have given a conceptual proof for general finite volume groups in the framework of representation theory. A far-reaching generalization to $\Gamma \subset G$, where G is a real rank 1 semisimple Lie group and Γ its discrete subgroup, was established by R. Miatello and N. Wallach [Mi-Wa]. A. Good pursued the study in the direction of Kloosterman sums associated with arbitrary one-parameter subgroups of a Fuchsian group in place of the stability groups of cusps. His work [Go] deserves further analysis.

From Theorem 9.3 one can quickly derive, using nothing more than the trivial estimate for Kloosterman sums (2.38), the following asymptotic formula:

$$(9.13) \qquad \sum_{|t_j|<T} |\nu_{aj}(n)|^2 + \sum_{\mathfrak{c}} \frac{1}{4\pi} \int_{-T}^{T} |\eta_{a\mathfrak{c}}(n,t)|^2 \, dt = \frac{2}{\pi} T^2 + O(|n| T)$$

for $n \neq 0$ and $T \geq 1$, where the implied constant depends on the group.

9.4. Kloosterman sums formula.

The formula (9.12) serves dual purposes. First it gives a tool for studying the spectrum as well as the Fourier coefficients of Maass forms on $\Gamma \backslash \mathbb{H}$ by means of Kloosterman sums. Through these sums one can employ a variety of methods beyond soft analysis; and when the group is arithmetic, one can use deep facts from algebraic geometry (Weil's bound for Kloosterman sums). The result is often sharper than what is possible for general groups. Next, assuming the spectral aspects are well understood, one can reverse the target to study sums of Kloosterman sums. At first glance this switch may seem to be useless; however, vast research in the early eighties has made this interplay very productive, due to new inputs from analytic number theory (see the surveys [Iw1], [Iw2]).

When applying (9.12) to study sums of Kloosterman sums, one desires a general test function on that side rather than on its spectral side. This leads us to the problem of reversing the transforms $h \mapsto h^-$ and $h \mapsto h^+$. If $mn < 0$, it is possible to solve this problem entirely by means of the Kontorovich-Lebedev inversion (1.31). Indeed, any function of class $\mathcal{C}^2$ on $[0, +\infty)$ satisfying the conditions

$$(9.14) \qquad f(0) = 0, \qquad f^{(j)}(x) \ll (x+1)^{-2-\varepsilon}, \qquad j = 0, 1, 2,$$

is realized in the image of the transform $h \mapsto h^-$; precisely, we have

$$f(x) = \frac{4}{\pi} \int_{-\infty}^{+\infty} K_{2it}(x) \, K_f(t) \, \sinh(\pi t) \, t \, dt,$$

where

$$(9.15) \qquad K_f(t) = \frac{4}{\pi} \cosh(\pi t) \int_{0}^{+\infty} K_{2it}(x) \, f(x) \, x^{-1} \, dx.$$

Then Theorem 9.3 for $h(t) = K_f(t)$ becomes

Theorem 9.4. *Let* $mn < 0$. *For any* $f(x)$ *satisfying* (9.14) *we have*

$$
\sum_c c^{-1} \mathcal{S}_{\mathfrak{ab}}(m, n; c) f\left(\frac{4\pi\sqrt{|mn|}}{c}\right)
$$

(9.16)
$$
= \sum_j K_f(t_j)\, \bar{\nu}_{\mathfrak{a}j}(m)\, \nu_{\mathfrak{b}j}(n)
$$

$$
+ \sum_{\mathfrak{c}} \frac{1}{4\pi} \int_{-\infty}^{\infty} K_f(t)\, \bar{\eta}_{\mathfrak{ac}}(m, t)\, \eta_{\mathfrak{bc}}(n, t)\, dt,
$$

where K_f *is the integral transform of* f *given by* (9.15).

If $mn > 0$, we need to invert $h \mapsto h^+$. But we are not yet fully equipped to perform this operation, because the image of this transform is not dense in the space of functions of main interest, specifically in $\mathfrak{L}^2(\mathbb{R}^+, x^{-1}\, dx)$. For a closer examination of this subspace let us write (9.10) as follows:

$$
h^+(x) = 4 \int_0^{+\infty} B_{2it}(x)\, h(t)\, \tanh(\pi t)\, t\, dt,
$$

where

(9.17)
$$
B_\nu(x) = \left(2 \sin \frac{\pi}{2}\nu\right)^{-1} \left(J_{-\nu}(x) - J_\nu(x)\right).
$$

Note that $B_{2it}(x) \in \mathfrak{L}^2(\mathbb{R}^+, x^{-1}\, dx)$. Therefore, the image of the transform $h \mapsto h^+$ falls into the subspace spanned by the functions $B_{2it}(x)$, $t \in \mathbb{R}$. It is a very large subspace, but not dense in $\mathfrak{L}^2(\mathbb{R}^+, x^{-1}dx)$. Indeed, the Bessel functions $J_\ell(x)$ of order $\ell \geq 1$, $\ell \equiv 1 \pmod{2}$, are missed, because they are orthogonal to $B_{2it}(x)$ (see Appendix B.5). Given $f \in \mathfrak{L}^2(\mathbb{R}^+, x^{-1}\, dx)$, put

(9.18)
$$
T_f(t) = \int_0^{+\infty} B_{2it}(x)\, f(x)\, x^{-1}\, dx,
$$

so $B_{2it}(x)\, T_f(t)$ is the projection of $f(x)$ onto $B_{2it}(x)$. Then define the continuous superposition of these projections by

(9.19)
$$
f^\infty(x) = 4 \int_0^{+\infty} B_{2it}(x)\, T_f(t)\, \tanh(\pi t)\, t\, dt.
$$

We also define

(9.20)
$$
f^\infty = \frac{2}{\pi} \int_0^{\infty} T_f(t)\, \tanh(\pi t)\, t\, dt.
$$

Incidentally, by (B.37) and (9.18) you can derive another formula for f^∞ which involves f directly, namely

$$(9.21) \qquad f^\infty = \frac{1}{2\pi} \int_0^\infty J_0(x) \, f(x) \, dx \, .$$

Then for $h(t) = T_f(t)$ Theorem 9.3 becomes

Theorem 9.5. *Let $mn > 0$. For any f satisfying (9.14) we have*

$$
\begin{aligned}
(9.22) \qquad \delta_{\mathfrak{ab}} \, \delta_{mn} \, f^\infty &+ \sum_c c^{-1} S_{\mathfrak{ab}}(m,n;c) \, f^\infty\left(\frac{4\pi\sqrt{mn}}{c}\right) \\
&= \sum_j T_f(t_j) \, \bar\nu_{\mathfrak{a}j}(m) \, \nu_{\mathfrak{b}j}(n) \\
&\quad + \sum_{\mathfrak{c}} \frac{1}{4\pi} \int_{-\infty}^\infty T_f(t) \, \bar\eta_{\mathfrak{ac}}(m,t) \, \eta_{\mathfrak{bc}}(n,t) \, dt \, .
\end{aligned}
$$

Although the functions of type $f^\infty(x)$ span most of the space which we often need when working with Kloosterman sums, they are never nice to deal with in practice; therefore it is important to determine the complementary function to $f^\infty(x)$. This is best described in terms of the Hankel transform

$$(9.23) \qquad H_f(x) = \int_0^{+\infty} J_0(xy) \, f(y) \, dy \, .$$

To recover f, use the Hankel inversion (see Appendix B.5)

$$(9.24) \qquad f(x) = \int_0^{+\infty} xy \, J_0(xy) \, H_f(y) \, dy \, .$$

One can prove under the conditions (9.14) that (see Appendix B.5)

$$(9.25) \qquad f^\infty(x) = \int_1^{+\infty} xy \, J_0(xy) \, H_f(y) \, dy \, ;$$

hence the complementary function $f^0(x) = f(x) - f^\infty(x)$ is given by

$$(9.26) \qquad f^0(x) = \int_0^1 xy \, J_0(xy) \, H_f(y) \, dy.$$

This said, we wish to have a formula complementary to (9.22) with $f^0(x)$ in place of $f^\infty(x)$. To find it, first recognize that $f^0(x)$ is the projection of

$f(x)$ onto the subspace of $\mathcal{L}^2(\mathbb{R}^+, x^{-1}dx)$ spanned by the Bessel functions $J_\ell(x)$ of order $\ell \geq 1$, $\ell \equiv 1 \pmod 2$, *i.e.* $f^0(x)$ is given by the Neumann series (see Appendix B.5)

$$(9.27) \qquad f^0(x) = \sum_{1 \leq \ell \equiv 1 \ (\text{mod } 2)} 2\ell \, N_f(\ell) \, J_\ell(x) \,,$$

where

$$(9.28) \qquad N_f(\ell) = \int_0^{+\infty} J_\ell(x) \, f(x) \, x^{-1} \, dx \,.$$

Note that $J_\ell(x) = i^{\ell+1} B_\ell(x)$; hence $N_f(\ell) = i^{\ell+1} T_f(i\ell/2)$. By virtue of the Neumann series expansion for $f^0(x)$, our problem reduces to finding an analogue of (9.12) with $J_\ell(x)$ as the test function attached to the Kloosterman sums. The latter will bring us to the classical (holomorphic) cusp forms of weight $k = \ell + 1$.

9.5. Petersson's formulas.

Let $\mathcal{M}_k(\Gamma)$ denote the linear space of holomorphic functions $f : \mathbb{H} \longrightarrow \mathbb{C}$ satisfying the following transformation rule:

$$(9.29) \qquad j_\gamma(z)^{-k} f(\gamma z) = f(z) \,, \qquad \gamma \in \Gamma \,.$$

These are called *automorphic forms of weight k with respect to the group Γ*. Throughout, we asume that k is even and positive. The space $\mathcal{M}_k(\Gamma)$ has finite dimension; precisely,

$$(9.30) \qquad \dim \mathcal{M}_k(\Gamma) = (k-1)(g-1) + \sum_{j=1}^{\ell} \left[\left(1 - \frac{1}{m_j}\right) \frac{k}{2} \right] + \frac{hk}{2}$$

if $k > 2$, where $(g; m_1, \ldots, m_\ell; h)$ is the signature of Γ (for $k = 2$ the formula is slighty different). Hence

$$\dim \mathcal{M}_k(\Gamma) \leq \frac{|F|}{4\pi} k + 1$$

by the Gauss-Bonnet formula (2.7).

Any $f \in \mathcal{M}_k(\Gamma)$ has the Fourier expansion in cusps of type

$$(9.31) \qquad j_{\sigma_\mathfrak{a}}(z)^{-k} f(\sigma_\mathfrak{a} z) = \sum_{n=0}^{\infty} \hat{f}_\mathfrak{a}(n) \, e(nz),$$

which converges absolutely and uniformly on compacta. If for every cusp
we have

(9.32) $$\hat{f}_{\mathfrak{a}}(0) = 0 \,,$$

then f is called a *cusp form*. A cusp form has exponential decay at cusps.
In particular, $y^{k/2} f(z)$ is bounded on $\mathbb{H}$. The subspace of cusp forms, say
$\mathcal{S}_k(\Gamma)$, is equipped with the Petersson inner product

(9.33) $$\langle f, g \rangle_k = \int_F y^k f(z) \, \bar{g}(z) \, d\mu z \,.$$

Observe that $y^k f(z) \, \bar{g}(z) \in \mathcal{A}(\Gamma)$, so the integral does not depend on the
choice of the fundamental domain.

As in the proof of Theorem 3.2, one shows that the Fourier coefficients
of a cusp form f normalized by $\langle f, f \rangle_k = 1$ satisfy the bound

(9.34) $$\sum_{1 \le n \le N} \frac{(k-1)!}{(4\pi n)^{k-1}} |\hat{f}_{\mathfrak{a}}(n)|^2 \ll c_{\mathfrak{a}}^{-1} N + k,$$

where the implied constant is absolute.

The space $\mathcal{S}_k(\Gamma)$ is spanned by the Poincaré series

(9.35) $$P_{\mathfrak{a}m}(z) = \sum_{\gamma \in \Gamma_{\mathfrak{a}} \backslash \Gamma} j_{\sigma_{\mathfrak{a}}^{-1}\gamma}(z)^{-k} \, e(m \, \sigma_{\mathfrak{a}}^{-1} \gamma z)$$

with $m \ge 1$. This is obvious by the following formula of Petersson:

(9.36) $$\langle f, P_{\mathfrak{a}m} \rangle_k = \frac{(k-2)!}{(4\pi m)^{k-1}} \, \hat{f}_{\mathfrak{a}}(m),$$

which is derived by the unfolding technique. Indeed, the subspace spanned
by the Poincaré series is closed (because $\mathcal{M}_k(\Gamma)$ has finite dimension), and
any function orthogonal to this subspace is zero by (9.36). Let us choose an
orthonormal basis of $\mathcal{S}_k(\Gamma)$, say $\{f_{jk}\}$, and expand $P_{\mathfrak{a}m}$ into this basis. By
(9.36) we get

(9.37) $$P_{\mathfrak{a}m}(z) = \frac{(k-2)!}{(4\pi m)^{k-1}} \sum_j \overline{\hat{f}_{\mathfrak{a}jk}(m)} \, f_{jk}(z) \,,$$

where $\hat{f}_{\mathfrak{a}jk}(m)$ denotes the m-th Fourier coefficient of f_{jk} in cusp $\mathfrak{a}$. On
the other hand, we have the Fourier expansion of $P_{\mathfrak{a}m}(z)$ in cusp $\mathfrak{b}$ due to
Petersson, Rankin and Selberg:

(9.38) $$j_{\sigma_{\mathfrak{b}}}(z)^{-k} P_{\mathfrak{a}m}(\sigma_{\mathfrak{b}} z) = \sum_{n=1}^{\infty} \left(\frac{n}{m} \right)^{(k-1)/2} \hat{P}_{\mathfrak{a}\mathfrak{b}}(m, n) \, e(nz) \,,$$

say, with

$$(9.39) \quad \hat{P}_{\mathfrak{ab}}(m,n) = \delta_{\mathfrak{ab}}\, \delta_{mn} + 2\pi i^k \sum_c c^{-1} \mathcal{S}_{\mathfrak{ab}}(m,n;c)\, J_{k-1}\left(\frac{4\pi\sqrt{mn}}{c}\right)$$

(for a proof, apply the double coset decomposition (2.21) as in the case of $E_{\mathfrak{a}m}(\sigma_{\mathfrak{b}}z|\psi)$ in Section 3.4). Comparing this with the n-th coefficient on the right side of (9.37), we arrive at the following:

Theorem 9.6. *Let m, n be positive integers and let k be a positive even integer. Then*

$$(9.40)$$
$$\frac{(k-2)!}{(4\pi\sqrt{mn})^{k-1}} \sum_j \overline{\hat{f}_{\mathfrak{a}jk}}(m)\, f_{\mathfrak{b}jk}(n)$$
$$= \delta_{\mathfrak{ab}}\, \delta_{mn} + 2\pi i^k \sum_c c^{-1} \mathcal{S}_{\mathfrak{ab}}(m,n;c)\, J_{k-1}\left(\frac{4\pi\sqrt{mn}}{c}\right).$$

Remarks. For $k = 2$ the Poincaré series (9.35) does not converge absolutely (a proper definition in this case was given by Hecke); nevertheless, (9.36)-(9.40) are true. The series of Kloosterman sums converges absolutely if $k \geq 4$ and at least conditionally if $k = 2$ by the spectral theory of Kloosterman sums, which we have already established.

Finally, we generalize (9.40) by summing in accordance with (9.27) (use (9.34) to validate the convergence). We obtain

Theorem 9.7. *Let $m, n > 0$. For any f satisfying (9.14) we have*

$$(9.41)$$
$$-\delta_{\mathfrak{ab}}\, \delta_{mn}\, f^{\infty} + \sum_c c^{-1} \mathcal{S}_{\mathfrak{ab}}(m,n;c)\, f^0\left(\frac{4\pi\sqrt{mn}}{c}\right)$$
$$= \sum_{2 \leq k \equiv 0 \;(\mathrm{mod}\;2)} \sum_j i^k\, N_f(k-1)\, \bar{\psi}_{\mathfrak{a}jk}(m)\, \psi_{\mathfrak{b}jk}(n),$$

where we have scaled the Fourier coefficients of an orthonormal basis $\{f_{jk}\}$ of $\mathcal{S}_k(\Gamma)$ down to

$$(9.42) \qquad \psi_{\mathfrak{a}jk}(m) = \left(\frac{\pi^{-k}\,\Gamma(k)}{(4m)^{k-1}}\right)^{1/2} \hat{f}_{\mathfrak{a}jk}(m).$$

All terms in (9.41) are obtained straightforwardly except for $-f^\infty$, which needs a few words of explanation. This term comes out as follows:

$$\sum_{1 \le \ell \equiv 1 \ (\text{mod } 2)} (2\pi i^k)^{-1} 2\ell \, N_f(\ell)$$

$$= \frac{1}{\pi} \int_0^{+\infty} \left(\sum_{r=1}^{\infty} (-1)^r (2r-1) \, J_{2r-1}(x) \right) f(x) \, x^{-1} \, dx$$

$$= \frac{-1}{2\pi} \int_0^{+\infty} J_0(x) \, f(x) \, dx = -f^\infty$$

by applying the recurrence relation $J_{n-1}(x) + J_{n+1}(x) = 2ny^{-1} J_n(x)$.

Theorem 9.7 constitutes the exact complement to Theorem 9.5. Adding (9.41) to (9.22), one gets a complete spectral decomposition for sums of Kloosterman sums. Notice that the diagonal terms $\delta_{\mathfrak{ab}} \delta_{mn} f^\infty$ cancel out, whereas the test functions attached to the Kloosterman sums make up the original function $f(x) = f^0(x) + f^\infty(x)$.

Theorem 9.8. *Let $m, n > 0$. For any f satisfying (9.14) we have*

(9.43)

$$\sum_c c^{-1} \mathcal{S}_{\mathfrak{ab}}(m, n; c) \, f\left(\frac{4\pi \sqrt{mn}}{c} \right)$$

$$= \sum_j T_f(t_j) \, \bar{\nu}_{\mathfrak{a}j}(m) \, \nu_{\mathfrak{b}j}(n)$$

$$+ \sum_{\mathfrak{c}} \frac{1}{4\pi} \int_{-\infty}^{\infty} T_f(t) \, \bar{\eta}_{\mathfrak{ac}}(m, t) \, \eta_{\mathfrak{bc}}(n, t) \, dt$$

$$+ \sum_{2 \le k \equiv 0 \ (\text{mod } 2)} \sum_j i^k \, N_f(k-1) \, \bar{\psi}_{\mathfrak{a}jk}(m) \, \psi_{\mathfrak{b}jk}(n) \, .$$

Remarks. On the spectral sides of (9.43) the Fourier coefficients of basic automorphic forms are multiplied by the integral transforms $i^k N_f(k-1) = T_f(i(k-1)/2)$ and $T_f(t_j)$ respectively (see Appendix B.5, where we call these the Neumann and the Titchmarsh coefficients of f, respectively). The partition $f(x) = f^0(x) + f^\infty(x)$ together with the series (9.27) and the integral (9.20) constitutes an inversion formula first established rigorously by D. B. Sears and E. C. Titchmarsh (see (4.6) of [Se-Ti, p.172]). A full characterization of functions which can be represented exclusively by the Neumann series (9.27) was given by G. H. Hardy and E. C. Titchmarsh [Ha-Ti].

Chapter 10

The Trace Formula

10.1. Introduction.

A truly beautiful formula has been derived from the spectral theorem by A. Selberg (see [Se2]). The Selberg trace formula establishes a quantitative connection between the spectrum and the geometry of the Riemann surface $\Gamma \backslash \mathbb{H}$.

A function $K : F \times F \longrightarrow \mathbb{C}$ and the integral operator having K as its kernel are said to be *of trace class* if $K(z, w)$ is absolutely integrable on the diagonal $z = w$, in which case the integral

$$\operatorname{Tr} K = \int_F K(z, z) \, d\mu z$$

is called the *trace*. Suppose for a moment that $\Gamma \backslash \mathbb{H}$ is a compact quotient and $K(z, w)$ is given by a smooth compactly supported function $k(u)$. Then K is of trace class. Integrating the spectral decomposition

$$K(z, z) = \sum_j h(t_j) \, |u_j(z)|^2,$$

we get

$$\operatorname{Tr} K = \sum_j h(t_j)$$

(the spectral trace of K, so to speak). On the other hand, from the series

$$K(z, z) = \sum_{\gamma \in \Gamma} k(z, \gamma z)$$

(the geometric side of K, so to speak), we get

$$\text{Tr}\, K = \sum_{\gamma \in \Gamma} \int_F k(z, \gamma z)\, d\mu z.$$

Comparing these two numbers, one obtains the pre-trace formula. It is a quite useful expression, yet it does not reveal the geometry of the surface $\Gamma \backslash \mathbb{H}$.

Following Selberg, we partition the group into conjugacy classes:

$$[\gamma] = \{\tau^{-1}\gamma\tau : \ \tau \in \Gamma\}.$$

Given a conjugacy class $\mathcal{C}$ in Γ, let $K_{\mathcal{C}}$ denote the partial kernel restricted to elements in $\mathcal{C}$,

$$K_{\mathcal{C}}(z, z) = \sum_{\gamma \in \mathcal{C}} k(z, \gamma z)\,.$$

Thus K and its trace split into

$$K = \sum_{\mathcal{C}} K_{\mathcal{C}}\,, \qquad \text{Tr}\, K = \sum_{\mathcal{C}} \text{Tr}\, K_{\mathcal{C}},$$

where

$$\text{Tr}\, K_{\mathcal{C}} = \sum_{\gamma \in \mathcal{C}} \int_F k(z, \gamma z)\, d\mu z\,.$$

Two elements $\tau, \tau' \in \Gamma$ yield the same conjugate of γ if and only if $\tau'\tau^{-1}$ belongs to the centralizer

$$Z(\gamma) = \{\rho \in \Gamma : \ \rho\gamma = \gamma\rho\}\,.$$

Therefore, we can write (choose $\gamma \in \mathcal{C}$)

$$\text{Tr}\, K_{\mathcal{C}} = \sum_{\tau \in Z(\gamma) \backslash \Gamma} \int_F k(z, \tau^{-1}\gamma\tau z)\, d\mu z$$

$$= \int_{Z(\gamma) \backslash \mathbb{H}} k(z, \gamma z)\, d\mu z,$$

where $Z(\gamma) \backslash \mathbb{H}$ is a fundamental domain of the centralizer, the point being that it is a relatively simple domain. Observe that the above integral really depends only on the conjugacy class of γ in the group $G = SL_2(\mathbb{R})$. Precisely, if $\gamma' = g^{-1}\gamma g$ with $g \in G$, then

$$\text{Tr}\, K_{\mathcal{C}} = \int_{g^{-1}Z(\gamma)g \backslash \mathbb{H}} k(z, g^{-1}\gamma g z)\, d\mu z = \text{Tr}\, K_{\mathcal{C}'},$$

where $g^{-1}Z(\gamma)g$ is the centralizer of γ' in $\Gamma' = g^{-1}\Gamma g$ and $\mathcal{C}' = g^{-1}\mathcal{C}g$. In particular, after conjugating Γ with suitable g, we can find a representative of the class in one of the groups $\pm N$, $\pm A$, K according to whether the class is parabolic, hyperbolic or elliptic (remember that every element of Γ has two representations in $G = SL_2(\mathbb{R})$). This special representative is convenient for computations, but above all it illuminates the geometric side of the trace.

Recall the classification of motions described in Section 1.5 and other facts about commutativity, fixed points, centralizers. We need to capture the conjugation representative of a motion in one of the groups $\pm N$, $\pm A$, K, which task leads us to examine the sets of fixed points. Looking from the surface $\Gamma\backslash\mathbb{H}$, we shall speak of equivalent classes of fixed points modulo Γ to refer to the fixed points of a whole conjugacy class rather than to these of a solitary motion. It does happen that two distinct conjugacy classes in Γ have the same fixed points modulo Γ.

The set of fixed points of a given $\gamma \neq \pm 1$ (the identity motion $\gamma = \pm 1$ having all $\mathbb{H}$ as fixed points is an exception; it has to be considered differently) determines the primitive class, say $\mathcal{C}_0$, and every other class $\mathcal{C}$ which has the same fixed points modulo Γ is a unique power of $\mathcal{C}_0$, say $\mathcal{C} = \mathcal{C}_0^\ell$, with $\ell \in \mathbb{Z}$, $\ell \neq 0$, with $1 \leq \ell < m$ if $\mathcal{C}_0$ is elliptic of order m. If $\gamma = \gamma_0^\ell$, where γ_0 is primitive, then $Z(\gamma) = Z(\gamma_0)$ is cyclic generated by γ_0; therefore the fundamental domain $Z(\gamma)\backslash\mathbb{H}$ is as simple as a vertical strip, a horizontal strip or a sector in $\mathbb{H}$, after γ is brought to $\pm N$, $\pm A$, K, respectively, by conjugation.

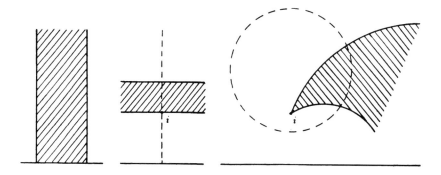

Figure 10. Fundamental domains of centralizers.

Our introductory arguments for derivation of the trace formula were oversimplified, since the convergence questions were ignored. A considerable difficulty occurs when the surface $\Gamma\backslash\mathbb{H}$ is not compact; thus, it has cusps which produce the continuous spectrum. In this case K is not of trace class on F for two parallel reasons: on the spectral side the Eisenstein series

$E_{\mathfrak{a}}(z, 1/2 + it)$ are not square integrable, whereas on the geometric side the partial kernels $K_{\mathcal{C}}(z, z)$ for parabolic classes are not absolutely integrable over cuspidal zones. Selberg has dealt with the problem by computing the trace asymptotically on the central part $F(Y) \subset F$ with Y tending to $+\infty$. Let $\mathrm{Tr}^Y K$ stand for such a truncated trace:

$$\mathrm{Tr}^Y K = \int_{F(Y)} K(z, z)\, d\mu z\,.$$

From the spectral side we obtain that $\mathrm{Tr}^Y K \sim A_1 \log Y + T_1$, whereas from the geometric side $\mathrm{Tr}^Y K \sim A_2 \log Y + T_2$, where A_1, A_2, T_1, T_2 are constants which can be explicitly expressed in terms of Γ. Hence one infers that $T_1 = T_2$, which is the celebrated trace formula (not a tautology like the relation $A_1 = A_2$).

In these lectures we apply Selberg's ideas to the iterated resolvent (4.14) given by the kernel

$$K(z, w) = G_s(z/w) - G_a(z/w),$$

where $G_s(z/w)$ is the automorphic Green function. Therefore the generating function

$$(10.1) \qquad\qquad k(u) = G_s(u) - G_a(u)$$

is smooth and bounded, but not compactly supported in $\mathbb{R}^+$, in contrast to our previous practice. In this context, let us record that the Selberg/Harish-Chandra transform of $k(u)$ is (use Theorems 1.14-1.17)

$$(10.2) \qquad h(r) = \left(\left(s - \frac{1}{2} \right)^2 + r^2 \right)^{-1} - \left(\left(a - \frac{1}{2} \right)^2 + r^2 \right)^{-1},$$

of which the Fourier transform is

$$(10.3) \qquad g(x) = (2s - 1)^{-1} e^{-|x|(s-1/2)} - (2a - 1)^{-1} e^{-|x|(a-1/2)}\,.$$

As a matter of fact, we shall carry out computations of particular components of the kernel for any k, h, g which satisfy adequate growth conditions. Yet, to be precise we cannot utilize these general computations until the convergence of many series and integrals involving the spectrum and the length of closed geodesics is established. In order to proceed without extra work, we are going to stick in the beginning to the particular functions given above with $a > s > 1$ so that k, h, g are positive and small,

$$(10.4) \qquad\qquad 0 < k(u) \ll (u + 1)^{-s}\,,$$

$$(10.5) \qquad\qquad 0 < h(r) \ll (|r| + 1)^{-4}\,,$$

$$(10.6) \qquad\qquad 0 < g(x) \ll e^{-|x|/2}\,.$$

Hence both series for $K(z, z)$, namely

$$(10.7) \qquad \sum_{\gamma \in \Gamma} k(u(z, \gamma z))$$

and

$$(10.8) \qquad \sum_{j} h(t_j) \left| u_j(z) \right|^2 + \sum_{\mathfrak{a}} \frac{1}{4\pi} \int_{-\infty}^{+\infty} h(r) \left| E_{\mathfrak{a}} \left(z, \frac{1}{2} + ir \right) \right|^2 dr,$$

converge absolutely and uniformly on compacta.

As a by-product of working with these particular test functions, on the side we shall release sufficient estimates to validate the convergence in the general case. From this point on we no longer assume the special case. Nevertheless we do continue looking after the resolvent kernel, because it will be fundamental in the theory of Selberg's zeta-function.

After the trace formula is established for the iterated resolvent and the required convergence is not a problem, we shall relax the condition $a > s > 1$ by analytic continuation. Then we recommend to the reader to generalize the formula by contour integration. Of course, the resulting integrals will be the same as those previously developed, however receiving them through another channel is an attractive exercise.

Governed by different goals, in some cases we shall elaborate more than one expression for the same thing. We shall have only the Fourier pair g, h present in the final formulation of the trace formula. The original kernel function k will be hiding behind the Selberg/Harish-Chandra transform.

10.2. Computing the spectral trace.

Integrating (10.8) over $F(Y)$ termwise, we get

$$\mathrm{Tr}^Y K = \sum_{j} h(t_j) \int_{F(Y)} |u_j(z)|^2 \, d\mu z$$

$$+ \frac{1}{4\pi} \int_{-\infty}^{+\infty} h(r) \sum_{\mathfrak{c}} \int_{F(Y)} \left| E_{\mathfrak{c}}^Y \left(z, \frac{1}{2} + ir \right) \right|^2 d\mu z \, dr \,.$$

Notice that we have changed the Eisenstein series into the truncated ones, because they match inside $F(Y)$.

Extending the integration to the whole of F, we immediately get an upper bound. For each point t_j in the discrete spectrum we get 1, and for each r in the continuous spectrum we get, by (6.35),

$$\sum_{\mathfrak{c}} \int_F |E_{\mathfrak{c}}^Y(z, \tfrac{1}{2}+ir)|^2 \, d\mu z = \operatorname{Tr} \langle \mathcal{E}^Y(\cdot, \tfrac{1}{2}+ir), {}^t\mathcal{E}^Y(\cdot, \tfrac{1}{2}+ir) \rangle$$

(10.9)
$$= \operatorname{Tr} (2ir)^{-1} \big(\Phi(\tfrac{1}{2}-ir) \, Y^{2ir} - \Phi(\tfrac{1}{2}+ir) \, Y^{-2ir} \big)$$

$$+ 2h \, \log Y - \frac{\varphi'}{\varphi}(\tfrac{1}{2}+ir) \, .$$

Here $\Phi(s)$ is the scattering matrix, h is its rank, *i.e.* the number of inequivalent cusps, $\varphi(s) = \det \Phi(s)$ and

$$\frac{\varphi'}{\varphi}(s) = \operatorname{Tr} \Phi'(s) \, \Phi^{-1}(s) \, .$$

For a proof of the last equation, employ the eigenvalues of $\Phi(s)$ and a unitary diagonalization. It follows from (10.9) that $-(\varphi'/\varphi)(1/2 + ir)$ is real and bounded below by a constant depending on the group. A good upper bound is not known, but see (10.13) below.

By (10.9) we need to evaluate the integral

$$I(Y) = \frac{1}{4\pi} \int_{-\infty}^{+\infty} \frac{h(r)}{2ir} \big(\Phi(\tfrac{1}{2}-ir) \, Y^{2ir} - \Phi(\tfrac{1}{2}+ir) \, Y^{-2ir} \big) \, dr \, .$$

To this end we write

$$I(Y) = \frac{1}{4\pi i} \int_{-\infty}^{+\infty} r^{-1} h(r) \big(\Phi(\tfrac{1}{2}-ir) \, Y^{2ir} - \Phi(\tfrac{1}{2}) \big) \, dr$$

by giving and taking back $\Phi(1/2)$ and then exploiting the symmetry $h(r) = h(-r)$. We now move the integration upwards to $\operatorname{Im} r = \varepsilon$, getting

$$I(Y) = -\Phi(\tfrac{1}{2}) \frac{1}{4\pi i} \int_{\operatorname{Im} r = \varepsilon} r^{-1} h(r) \, dr + O(Y^{-2\varepsilon}),$$

because $\Phi(s)$ is bounded in $1/2 \le \operatorname{Re} s \le 1/2 + \varepsilon$ for a small ε. Here

(10.10)
$$\frac{1}{2\pi i} \int_{\operatorname{Im} r = \varepsilon} r^{-1} h(r) \, dr = -\frac{1}{2} h(0) \, .$$

To see this, we move the integration downwards to $\operatorname{Im} r = -\varepsilon$, passing a simple pole at $r = 0$ of residue $h(0)$. By the symmetry $h(r) = h(-r)$ the lower horizontal line integral is equal to minus the upper one, which therefore is equal to half of the residue. We end up with

$$I(Y) = \frac{1}{4}\,\Phi(\frac{1}{2})\,h(0) + O(Y^{-2\varepsilon})\,.$$

Finally, summing $h(t_j)$ over the point spectrum as well as integrating other parts against $h(r)$, we conclude the following inequality for the truncated trace:

$$\text{(10.11)} \qquad \begin{aligned} \text{Tr}^{\,Y} K &< \sum_j h(t_j) + \frac{1}{4\pi}\int_{-\infty}^{+\infty} \frac{-\varphi'}{\varphi}(\frac{1}{2}+ir)\,h(r)\,dr \\ &\quad + \frac{1}{4}\,h(0)\,\text{Tr}\,\Phi(\frac{1}{2}) + g(0)\,h\,\log Y + O(Y^{-\varepsilon})\,. \end{aligned}$$

Here for the particular $k(u)$ given by (10.1) we have explicitly

$$g(0) = (2s-1)^{-1} - (2a-1)^{-1}\,,$$

$$\frac{1}{4}\,h(0) = (2s-1)^{-2} - (2a-1)^{-2}\,,$$

$$h(t_j) = \frac{1}{2s-1}\Big(\frac{1}{s-s_j} - \frac{1}{1-s-s_j}\Big) - \frac{1}{2a-1}\Big(\frac{1}{a-s_j} - \frac{1}{1-a-s_j}\Big)\,.$$

In order to get a lower bound for $\text{Tr}^{\,Y} K$, we must estimate and subtract from the upper bound (10.11) the truncated traces over cuspidal zones $F_{\mathfrak{a}}(Y)$. If u_j is a cusp form, we infer from (8.3) that

$$\int_{F_{\mathfrak{a}}(Y)} |u_j(z)|^2\,d\mu z \ll |s_j|\,Y^{-2}\,.$$

If u_j is the residue of an Eisenstein series at s_j with $1/2 < s_j \le 1$, we infer from (8.4) that

$$\int_{F_{\mathfrak{a}}(Y)} |u_j(z)|^2\,d\mu z \ll Y^{1-2s_j}\,.$$

From these two estimates we conclude using (7.11) that

$$\sum_j h(t_j)\int_{F\backslash F(Y)} |u_j(z)|^2\,d\mu z \ll Y^{-2\varepsilon}\,.$$

With the truncated Eisenstein series $E_{\mathfrak{c}}^{\,Y}(z,s)$ the argument is more involved. We use the inequality

$$\int_Y^{+\infty} |W_s(iy)|^2\,y^{-2}\,dy \ll |s| \int_{Y/2}^{+\infty} |W_s(iy)|^2\,y^{-3}\,dy$$

to estimate as follows:

$$\int\limits_{F_{\mathfrak{a}}(Y)} |E_{\mathfrak{c}}^Y(z,s)|^2 \, d\mu z = \sum_{n \neq 0} |\varphi_{\mathfrak{a}\mathfrak{c}}(n,s)|^2 \int_Y^{+\infty} |W_s(iny)|^2 \, y^{-2} \, dy$$

$$\ll |s| \int_{Y/2}^{+\infty} \sum_{n \neq 0} |\varphi_{\mathfrak{a}\mathfrak{c}}(n,s) \, W_s(iny)|^2 \, y^{-3} \, dy \, .$$

Hence, we infer from (8.24) that

$$(10.12) \qquad \int_{-R}^R \int\limits_{F_{\mathfrak{a}}(Y)} |E_{\mathfrak{c}}^Y(z, \tfrac{1}{2} + ir)|^2 \, d\mu z \, dr \ll R^3 \, Y^{-1} \, .$$

Then, combining with (10.5), we conclude that

$$\frac{1}{4\pi} \int_{-\infty}^{+\infty} h(r) \sum_{\mathfrak{c}} \int\limits_{F \backslash F(Y)} |E_{\mathfrak{c}}^Y(z, \tfrac{1}{2} + ir)|^2 \, d\mu z \, dr \ll Y^{-1} \, .$$

We have shown above that the truncated traces over cuspidal zones are absorbed by the error term in (10.11), so the inequality (10.11) turns into an equality.

As a by-product of the work done in this section we state the following estimate:

$$(10.13) \qquad M_\Gamma(R) = \frac{1}{4\pi} \int_{-R}^R \frac{-\varphi'}{\varphi}(\tfrac{1}{2} + ir) \, dr \ll R^2 \, .$$

To see this, we integrate (7.10) over $F(Y)$ with $Y \approx R$ and apply (10.9) together with (10.12).

10.3. Computing the trace for parabolic classes.

As indicated in the introduction, we shall compute the geometric traces for each conjugacy class separately. We begin with the parabolic motions, since they require special care. There are h primitive parabolic conjugacy classes, one for each class of equivalent cusps. The primitive class, say $\mathcal{C}_{\mathfrak{a}}$, attached to cusps equivalent with $\mathfrak{a}$ consists of generators of the stability groups of these cusps. Every parabolic conjugacy class is obtained as a power $\mathcal{C} = \mathcal{C}_{\mathfrak{a}}^\ell$ for some $\mathfrak{a}$ and $\ell \neq 0$. Let $\gamma = \gamma_{\mathfrak{a}}^\ell$, where $\gamma_{\mathfrak{a}}$ is the generator of $\Gamma_{\mathfrak{a}}$, so the centralizer $Z(\gamma) = Z(\gamma_{\mathfrak{a}}) = \Gamma_{\mathfrak{a}}$ is the stability group. By the unfolding technique, the truncated trace of the class $\mathcal{C}$ evolves into

$$\mathrm{Tr}^Y K_{\mathcal{C}} = \int\limits_{Z(\gamma) \backslash \mathbb{H}(Y)} k(z, \gamma z) \, d\mu z,$$

where $\mathbb{H}(Y)$ is the region of the upper half plane with the cuspidal zones removed. Conjugating by the scaling matrix $\sigma_\mathfrak{a}$, we get

$$\mathrm{Tr}^Y K_\mathcal{C} = \int\limits_{B\backslash\sigma_\mathfrak{a}\mathbb{H}(Y)} k(z, z+\ell)\,d\mu z\,.$$

Notice that the set $B\backslash\sigma_\mathfrak{a}\mathbb{H}(Y)$ is contained in the box $\{z : 0 < x \leq 1, 0 < y \leq Y\}$ and contains the box $\{z : 0 < x \leq 1, Y' < y \leq Y\}$, where $Y'Y = c_\mathfrak{a}^{-2}$. Therefore,

$$\int_0^1 \int_{Y'}^Y k(z, z+\ell)\,d\mu z \leq \mathrm{Tr}^Y K_\mathcal{C} \leq \int_0^1 \int_0^Y k(z, z+\ell)\,d\mu z\,.$$

Here we have

$$\int_0^1 \int_0^Y k(z, z+\ell)\,d\mu z = \int_0^Y k\big((\frac{\ell}{2y})^2\big)\, y^{-2}\,dy$$

$$= |\ell|^{-1} \int_{(\ell/2Y)^2}^{+\infty} k(u)\, u^{-1/2}\,du\,.$$

To continue the computation we first sum over ℓ, getting

$$2\int_{(2Y)^{-2}}^{+\infty} k(u)\, u^{-1/2}\Big(\sum_{1\leq\ell<2Y\sqrt{u}} \ell^{-1}\Big)\,du$$

$$= 2\int_{(2Y)^{-2}}^{+\infty} k(u)\, u^{-1/2}\big(\log 2Y\sqrt{u} + \gamma + O(u^{-1/2}\,Y^{-1})\big)\,du$$

$$= L(Y) + O(Y^{-1}\log Y),$$

where

$$L(Y) = 2\int_0^{+\infty} k(u)\, u^{-1/2}\big(\log 2Y\sqrt{u} + \gamma\big)\,du\,.$$

And with Y replaced by Y' we get $O(Y')$ by a simpler argument:

$$O(Y') \ll 2\int_{(2Y)^{-2}}^{+\infty} k(u)\, u^{-1/2}\log(u+2)\,du \ll Y'.$$

Hence we conclude that all parabolic motions having fixed points equivalent to $\mathfrak{a}$ yield

$$(10.14) \qquad \sum_{\mathcal{C}=\mathcal{C}_\mathfrak{a}^\ell} \mathrm{Tr}^Y K_\mathcal{C} = L(Y) + O(Y^{-1}\log Y)\,.$$

It remains to evaluate $L(Y)$. We split $L(Y)$ into

(10.15) $$L(Y) = g(0)\,(\log 2Y + \gamma) + \int_0^{+\infty} k(u)\,u^{-1/2}\log u\,du,$$

where the first term is obtained by (see (1.62))

$$\int_0^{+\infty} k(u)\,u^{-1/2}\,du = q(0) = \frac{1}{2}\,g(0)\,.$$

To the second term we apply (1.64), getting

$$\int_0^{+\infty} k(u)\,u^{-1/2}\log u\,du = \frac{-1}{\pi}\int_0^{+\infty}\left(\int_0^v \frac{\log u}{\sqrt{u(v-u)}}\,du\right)dq(v)$$

$$= \frac{-1}{\pi}\int_0^{+\infty}\left(\int_0^1 \frac{\log uv}{\sqrt{u(1-u)}}\,du\right)dq(v)$$

$$= \frac{1}{\pi}\,q(0)\int_0^1 \frac{\log u}{\sqrt{u(1-u)}}\,du - \frac{1}{\pi}\int_0^1 \frac{du}{\sqrt{u(1-u)}}\int_0^{+\infty}\log v\,dq(v)$$

(do not try to integrate by parts!). In the last line the first integral is $-2\pi\log 2$, the second is π and the third is

$$\int_0^{+\infty}\log v\,dq(v) = \int_0^{+\infty}\log(\sinh\frac{r}{2})\,dg(r),$$

upon changing v into $(\sinh(r/2))^2$. Collecting the above evaluations, we arrive at

(10.16) $$L(Y) = g(0)(\log Y + \gamma) - \int_0^{+\infty}\log(\sinh\frac{r}{2})\,dg(r)\,.$$

If one prefers to have an expression in terms of h rather than g, we supply the formula

(10.17)
$$\int_0^{+\infty}\log(\sinh\frac{r}{2})\,dg(r) = g(0)(\gamma + \log 2) - \frac{1}{4}\,h(0)$$
$$+ \frac{1}{2\pi}\int_{-\infty}^{+\infty} h(t)\,\psi(1+it)\,dt,$$

where

(10.18) $$\psi(s) = \frac{\Gamma'}{\Gamma}(s) = -\gamma - \sum_{n=0}^{\infty}\left(\frac{1}{n+s} - \frac{1}{n+1}\right)\,.$$

For the proof we write

$$g'(r) = -\frac{1}{2\pi i} \int\limits_{\mathrm{Im}\,t=\varepsilon} e^{irt} h(t)\,t\,dt$$

and employ the formula for the Laplace transform of $\log(\sinh r/2)$ (see [Gr-Ry, 4.331.1, p. 573, and 4.342.3, p. 575]):

(10.19) $$\int_0^{+\infty} \log(\sinh\frac{r}{2})\,de^{-\nu r} = \gamma + \log 2 - \frac{1}{2\nu} + \psi(1+\nu)\,.$$

With these ingredients one derives (10.17) straightforwardly, except for the term $-h(0)/4$, which comes from (10.10). Combining (10.17) and (10.16), we get

(10.20) $$L(Y) = g(0)\log\frac{Y}{2} + \frac{1}{4}h(0) - \frac{1}{2\pi}\int_{-\infty}^{+\infty} h(t)\,\psi(1+it)\,dt\,.$$

For the particular $g(r)$ given by (10.3) we get immediately from (10.16) and (10.19) that

(10.21) $$\begin{aligned} L(Y) = (2s-1)^{-1}(\log Y - \psi(s+\frac{1}{2}) - \log 2) + (2s-1)^{-2} \\ - \ (\text{the same for } s = a)\,. \end{aligned}$$

Before proceeding to the non-parabolic conjugacy classes, let us observe that the leading terms $g(0)\,h\,\log Y$ in (10.14) and (10.11) coincide. After subtracting $g(0)\,h\,\log Y$, what is left on the spectral side converges to a constant as $Y \to \infty$. This proves (by the positivity of $k(u)$ on the geometric side in case of Green's kernel, and in general by majorization) that the remaining partial kernels associated with the non-parabolic classes are of trace type, individually and in total. Therefore we can simplify the job by computing the actual traces which are the limits of the truncated ones.

10.4. Computing the trace for the identity motion.

The identity motion forms a class by itself, $\mathcal{C} = \{1\}$. Thus $K_{\mathcal{C}}(z,w) = k(z,w)$ and

(10.22) $$\mathrm{Tr}\,K_{\mathcal{C}} = \int_F k(z,z)\,d\mu z = k(0)\,|F|,$$

where $|F|$ is the area of a fundamental domain. By (1.64') we have

(10.23) $$k(0) = \frac{1}{4\pi}\int_{-\infty}^{+\infty} r\,\tanh(\pi r)\,h(r)\,dr\,.$$

For the particular $h(r)$ given in (10.2) we compute directly that

$$k(0) = \lim_{u \to 0} \big(G_s(u) - G_a(u) \big)$$

$$= \lim_{u \to 0} \frac{1}{4\pi} \int_0^1 \left(\left(\frac{\xi(1-\xi)}{\xi + u} \right)^{s-1} - \left(\frac{\xi(1-\xi)}{\xi + u} \right)^{a-1} \right) \frac{d\xi}{\xi + u}$$

(10.24)

$$= \frac{1}{4\pi} \int_0^1 \left((1-\xi)^{s-1} - (1-\xi)^{a-1} \right) \xi^{-1} \, d\xi$$

$$= \frac{1}{4\pi} \sum_{n=0}^{\infty} \left(\frac{1}{n+s} - \frac{1}{n+a} \right) = \frac{1}{4\pi} \big(\psi(a) - \psi(s) \big) .$$

10.5. Computing the trace for hyperbolic classes.

The primitive hyperbolic conjugacy classes in Γ are the most fascinating of all. Following Selberg, we denote such a class by P, displaying its resemblance to prime ideals in number fields. Let $\mathcal{C} = P^\ell$. Choose $\gamma_P \in P$ and $\gamma = \gamma_P^\ell \in \mathcal{C}$. Then $Z(\gamma) = Z(\gamma_P)$ and

$$\operatorname{Tr} K_{\mathcal{C}} = \int_{Z(\gamma_P) \backslash \mathbb{H}} k(z, \gamma z) \, d\mu z \, .$$

By conjugation in G we send γ_P to $\pm A$. The resulting motion acts simply as a dilation $z \mapsto pz$ by a positive factor $p \neq 1$. Suppose that $p > 1$, or else change P into P^{-1} (this only interchanges the fixed points of the classes $\mathcal{C} = P^\ell$). Then $\log p$ is the hyperbolic distance of i to pi, thus also the distance of z to $\gamma_P z$ for any z on the geodesic joining the fixed points of γ_P. Since γ_P maps the geodesic joining its two fixed points in $\hat{\mathbb{R}}$ into itself (not identically), this geodesic closes on the surface $\Gamma \backslash \mathbb{H}$ on which the points $z, \gamma_P z$ are the same. Thus $\log p$ is just the length of the closed geodesic multiplied by the winding number (the geodesic segment joining z with $\gamma_P z$ in the free space may wind itself on $\Gamma \backslash \mathbb{H}$ a finite number of times). We shall define $p = NP$ and call it the norm of P (it does not depend on representatives in the conjugacy class). The norm can be expressed in terms of the trace as

$$\operatorname{Tr} P = NP^{1/2} + NP^{-1/2} \, .$$

After conjugation in G, the centralizer becomes a cyclic group generated by the dilation $z \mapsto pz$, so its fundamental domain is the horizontal strip $1 < y < p$. Hence we obtain

$$\operatorname{Tr} K_{\mathcal{C}} = \int_1^p \int_{-\infty}^{+\infty} k(z, p^\ell z) \, d\mu z \, .$$

Furthermore, putting $2d = |p^{\ell/2} - p^{-\ell/2}|$, we continue the computation as follows:

$$
\begin{aligned}
\operatorname{Tr} K_{\mathcal{C}} &= \int_1^p \int_{-\infty}^{+\infty} k\big((\frac{d|z|}{y})^2\big)\, y^{-2}\, dx\, dy \\
&= \Big(\int_1^p y^{-1}\, dy\Big) \int_{-\infty}^{+\infty} k(d^2(x^2+1))\, dx \\
(10.25) \qquad &= \frac{\log p}{d} \int_{d^2}^{+\infty} \frac{k(u)}{\sqrt{u-d^2}}\, du = \frac{\log p}{d}\, q(d^2) \\
&= \frac{\log p}{2d}\, g\big(2\log(\sqrt{d^2+1}+d)\big) = \frac{\log p}{2d}\, g(\ell \log p) \\
&= |p^{\ell/2} - p^{-\ell/2}|^{-1} g(\ell \log p)\, \log p\,.
\end{aligned}
$$

In particular, for $k(u)$ given by (10.1) we get from (10.3) and (10.25) that the trace for the class $\mathcal{C} = P^\ell$ is

$$
(10.26) \qquad \operatorname{Tr} K_{\mathcal{C}} = (2s-1)^{-1}(1 - p^{-|\ell|})^{-1} p^{-|\ell|s} \log p \\
- \text{ (the same for } s = a)\,.
$$

Summing over $l > 0$, we get the total trace from all hyperbolic classes having equivalent fixed points:

$$
(10.27) \qquad \sum_{\mathcal{C} = P^\ell} \operatorname{Tr} K_{\mathcal{C}} = (2s-1)^{-1} \sum_{k=0}^{\infty} (p^{s+k} - 1)^{-1} \log p \\
- \text{ (the same for } s = a)\,.
$$

10.6. Computing the trace for elliptic classes.

The idea is the same, but computations are somewhat harder. We denote by $\mathcal{R}$ a primitive elliptic conjugacy class in Γ. There are only a finite number of these. Let $m = m_{\mathcal{R}} > 1$ be the order of $\mathcal{R}$. Any elliptic class having the same fixed points as $\mathcal{R}$ is $\mathcal{C} = \mathcal{R}^\ell$ with $0 < \ell < m$. Conjugating $\mathcal{R}$ in G, one can assume the representative to be $k(\theta)$, where $\theta = \theta_{\mathcal{R}} = \pi m^{-1}$; this acts as a rotation of angle 2θ at i. Since it generates the centralizer, the fundamental domain of that centralizer is a hyperbolic sector of angle 2θ at i, say $\mathcal{S}$. Therefore, we have

$$
\operatorname{Tr} K_{\mathcal{C}} = \int_{\mathcal{S}} k(z, k(\theta\ell)z)\, d\mu z = \frac{1}{m} \int_{\mathbb{H}} k(z, k(\theta\ell)z)\, d\mu z,
$$

because it takes m images of $\mathcal{S}$ to cover $\mathbb{H}$ exactly (except for a zero measure set). We shall continue computation in the geodesic polar coordinates $z =$

$k(\varphi)\,e^{-r}i$, where φ ranges over $[0,\pi)$ and r over $[0,+\infty)$ (see Section 1.3). Since $k(\theta\ell)$ commutes with $k(\varphi)$, we get

$$\operatorname{Tr} K_{\mathcal{C}} = \frac{\pi}{m}\int_0^{+\infty} k(e^{-r}i,\,k(\theta\ell)\,e^{-r}i)\,(2\sinh r)\,dr\,.$$

By $u(z,k(\theta)z) = (2y)^{-2}|z^2+1|^2 \sin^2\theta = (\sinh r\,\sin\theta)^2$ if $z = e^{-r}i$ we get several formulas for $\operatorname{Tr} K_{\mathcal{C}}$ in terms of $k(u)$:

$$\operatorname{Tr} K_{\mathcal{C}} = \frac{\pi}{m}\int_0^{+\infty} k\big((\sinh r\,\sin\theta\ell)^2\big)\,(2\sinh r)\,dr$$

(10.28)
$$= \frac{\pi}{m\,\sin\theta\ell}\int_0^{+\infty}\frac{k(u)\,du}{\sqrt{u+\sin^2\theta\ell}}$$

$$= \frac{\pi}{m}\int_0^{+\infty} k\big(u\,\sin^2\tfrac{\pi\ell}{m}\big)\,(u+1)^{-1/2}\,du\,.$$

These are nice and practical expressions; nevertheless, we continue computing, since our strategy is to remove k from the scene. Applying (1.64), by partial integration, we get (for $a > 0$)

$$\int_0^{+\infty} k(u)\,(u+a^2)^{-1/2}\,du$$

$$= -\frac{1}{\pi}\int_0^{+\infty} q'(v)\int_0^v \big((v-u)(u+a^2)\big)^{-1/2}\,du\,dv$$

$$= -\frac{1}{\pi}\int_0^{+\infty} q'(v)\int_0^{v/(v+a^2)} \big(u(1-u)\big)^{-1/2}\,du\,dv$$

$$= \frac{a}{\pi}\int_0^{+\infty} q(v)\,(v+a^2)^{-1}v^{-1/2}\,dv$$

$$= \frac{a}{2\pi}\int_0^{+\infty}\frac{g(r)\,\cosh(r/2)}{\sinh^2(r/2)+a^2}\,dr$$

by changing $v = \sinh^2(r/2)$. For $a = \sin\alpha > 0$ this yields

$$\int_0^{+\infty} k(u)\,(u+\sin^2\alpha)^{-1/2}\,du = \frac{\sin\alpha}{\pi}\int_0^{+\infty}\frac{g(r)\,\cosh(r/2)}{\cosh r - \cos 2\alpha}\,dr\,,$$

and taking $\alpha = \pi\ell m^{-1}$ we conclude that

(10.29)
$$\operatorname{Tr} K_{\mathcal{C}} = \frac{1}{m}\int_0^{+\infty}\frac{g(r)\,\cosh(r/2)}{\cosh r - \cos(2\pi\ell/m)}\,dr\,.$$

If one prefers to have an expression in terms of h rather than g, we state another formula (a proof is cumbersome):

(10.30)
$$\operatorname{Tr} K_{\mathcal{C}} = \big(2\,m\sin\tfrac{\pi\ell}{m}\big)^{-1}\int_{-\infty}^{+\infty} h(r)\,\frac{\cosh\pi(1-2\ell/m)r}{\cosh\pi r}\,dr\,.$$

Now let us apply (10.29) for g from (10.3). We begin by an appeal to the following formula:

$$\int_0^{+\infty} e^{-\mu x}(\cosh x - \cos t)^{-1}\,dx = \frac{2}{\sin t}\sum_{k=1}^{\infty}\frac{\sin kt}{\mu + k},$$

valid for $\operatorname{Re}\mu > -1$ and $t \neq 2\pi n$ (see [Gr-Ry, 3.543.2, p. 357]). Adding this up for the two values $\mu - 1/2$ and $\mu + 1/2$, we derive

$$\int_0^{+\infty} e^{-\mu r}\left(\cosh\frac{r}{2}\right)(\cosh r - \cos 2\alpha)^{-1}\,dr = \frac{1}{\sin\alpha}\sum_{k=0}^{\infty}\frac{\sin(2k+1)\alpha}{k + \mu + 1/2}\,.$$

This yields

$$\operatorname{Tr}K_{\mathcal{C}} = \left((2s-1)\,m\,\sin\frac{\pi\ell}{m}\right)^{-1}\sum_{k=0}^{\infty}(s+k)^{-1}\sin(2k+1)\frac{\pi\ell}{m}$$
$$-\ \text{(the same for } s = a\text{)}\,.$$

Next we exploit the periodicity to break the summation into residue classes modulo m as follows:

$$\sum_k = \sum_{0\le k<m}\sin(2k+1)\frac{\pi\ell}{m}\sum_{n=0}^{\infty}\left((s+k+mn)^{-1} - (m+mn)^{-1}\right).$$

Here we have borrowed the terms $(m + mn)^{-1}$ to produce convergence at no cost, because

$$\sum_{0\le k<m}\sin(2k+1)\frac{\pi\ell}{m} = 0\,.$$

By the same token we can borrow the Euler constant. We get (see (10.18))

$$\sum_k = \frac{-1}{m}\sum_{0\le k<m}\psi\left(\frac{s+k}{m}\right)\sin(2k+1)\frac{\pi\ell}{m}\,.$$

Hence, we arrive at

(10.31) $$\operatorname{Tr}K_{\mathcal{C}} = \frac{-1}{(2s-1)\,m^2}\sum_{0\le k<m}\psi\left(\frac{s+k}{m}\right)\frac{\sin(2k+1)\pi\ell/m}{\sin\pi\ell/m}$$
$$-\ \text{the same for } s = a\,.$$

Finally, we sum over $0 < \ell < m$ to compute the total trace of all elliptic classes $\mathcal{C} = \mathcal{R}^\ell$ which have common fixed points mod Γ. It follows from the identity

$$\sum_{|n|\le k} e\left(\frac{\ell n}{m}\right) = \frac{\sin(2k+1)\pi\ell/m}{\sin\pi\ell/m}$$

that

$$\sum_{0<\ell<m} \frac{\sin(2k+1)\pi\ell/m}{\sin \pi\ell/m} = m - 2\,k - 1\,.$$

Therefore,

$$(10.32) \qquad \sum_{\mathcal{C}=\mathcal{R}^\ell} \operatorname{Tr} K_{\mathcal{C}} = \frac{1}{(2s-1)\,m} \sum_{0\leq k<m} \left(\frac{2k+1}{m} - 1\right) \psi\!\left(\frac{s+k}{m}\right).$$

Another interesting transformation is offered by the identity

$$\frac{1}{m} \sum_{0\leq k<m} \psi\!\left(\frac{s+k}{m}\right) = \psi(s) - \log m$$

(see [Gr-Ry, 8.365.6, p. 945]). Writing

$$\frac{2k+1}{m} - 1 = \frac{2s+2k-m}{m} - \frac{2s-1}{m}\,,$$

it leads to

$$(10.33) \qquad \sum_{\mathcal{C}=\mathcal{R}^\ell} \operatorname{Tr} K_{\mathcal{C}} = m^{-1}\left(\log m - \psi(s)\right) + (2s-1)^{-1} R_m(s)\,,$$

where

$$(10.34) \qquad R_m(s) = m^{-2} \sum_{0\leq k<m} (2s+2k-m)\,\psi\!\left(\frac{s+k}{m}\right).$$

The key point in the last arrangement is that $R_m(s)$ is meromorphic in the whole complex s-plane with only simple poles at non-positive integers $-d$ of residue $2[d/m]+1$, which is a positive integer. These properties are vital for constructing Selberg's zeta-function.

10.7. Trace formulas.

All parts from which to build the trace formula have now been manufactured. Let us first assemble these for the particular pair h, g given by (10.2) and (10.3).

Theorem 10.1 (Resolvent Trace Formula). *Let $a > 1$ and $\operatorname{Re} s > 1$. We have*

(10.35)

$$
\sum_j \left(\frac{1}{(s-1/2)^2 + t_j^2} - \frac{1}{(a-1/2)^2 + t_j^2} \right)
$$

$$
+ \frac{1}{4\pi} \int_{-\infty}^{\infty} \left(\frac{1}{(s-1/2)^2 + r^2} - \frac{1}{(a-1/2)^2 + r^2} \right) \frac{-\varphi'}{\varphi} \left(\tfrac{1}{2} + ir \right) dr
$$

$$
= \frac{1}{(2s-1)^2} \left(h - \operatorname{Tr} \Phi\!\left(\tfrac{1}{2}\right) \right) - \frac{h}{2s-1} \left(\psi\!\left(s + \tfrac{1}{2}\right) + \log 2 \right)
$$

$$
- \psi(s) \frac{|F|}{2\pi} + \frac{1}{2s-1} \sum_P \sum_{k=0}^{\infty} \frac{\log p}{p^{s+k} - 1}
$$

$$
+ \frac{1}{2s-1} \sum_{\mathcal{R}} \sum_{0 \le k < m} \frac{2k+1-m}{m^2} \psi\!\left(\frac{s+k}{m}\right)
$$

$$
- \; (\text{the same for } s = a) .
$$

Remarks. In the above formula each term of the discrete spectrum is counted with the multiplicity of the eigenvalue $\lambda_j = t_j^2 + 1/4$, h is the number of primitive parabolic classes (= the number of inequivalent cusps), P ranges over primitive hyperbolic classes of norm $p = NP > 1$ and $\mathcal{R}$ ranges over primitive elliptic classes of order $m = m_{\mathcal{R}} > 1$. These terms come from (10.11), (10.21), (10.24), (10.27) and (10.32), respectively. A more general resolvent trace formula is given by J. Fischer [Fi].

The series over the discrete spectrum and the integral accounting for the continuous spectrum in the resolvent trace formula converge absolutely, due to (7.11) and (10.13). Therefore the Dirichlet series over the hyperbolic classes also converges absolutely in $\operatorname{Re} s > 1$; in fact, one gets quickly from the lowest eigenvalue $\lambda_0 = 0$ that

(10.36)
$$
\sum_P p^{-s} \log p \sim \frac{1}{s-1}, \qquad \text{as } s \to 1^+ .
$$

The above observations permit us to construct the trace formula for a general pair h, g.

Theorem 10.2 (Selberg's Trace Formula). *Suppose h satisfies the conditions (1.63), and let g be the Fourier transform of h. Then*

$$\sum_j h(r_j) + \frac{1}{4\pi} \int_{-\infty}^{\infty} h(r) \frac{-\varphi'}{\varphi}\left(\frac{1}{2}+ir\right) dr$$

(10.37)
$$= \frac{|F|}{4\pi} \int_{-\infty}^{\infty} h(r)\, r\, \tanh(\pi r)\, dr$$

$$+ \sum_P \sum_{\ell=1}^{\infty} (p^{\ell/2}-p^{-\ell/2})^{-1} g(\ell \log p) \log p$$

$$+ \sum_{\mathcal{R}} \sum_{0<\ell<m} \left(2\, m\, \sin\frac{\pi\ell}{m}\right)^{-1} \int_{-\infty}^{\infty} h(r) \frac{\cosh \pi(1-2\ell/m)r}{\cosh \pi r}\, dr$$

$$+ \frac{h(0)}{4} \operatorname{Tr}\left(I - \Phi\left(\frac{1}{2}\right)\right) - h\, g(0) \log 2$$

$$- \frac{h}{2\pi} \int_{-\infty}^{\infty} h(r)\, \psi(1+ir)\, dr\,.$$

Remarks. The above terms come from (10.11), (10.20), (10.23), (10.25) and (10.30). The series and integrals converge absolutely. For alternative expressions, see (10.16) and (10.29).

10.8. The Selberg zeta-function.

In connection with the trace formula, A. Selberg (see [Se2]) has introduced a zeta-function which in many ways mimics the L-functions of algebraic number fields. As in classical cases, the zeta-function is built with various local factors. We define

(10.38)
$$Z_\Gamma(s) = \prod_P \prod_{k=0}^{\infty} (1 - p^{-s-k}) \qquad \text{if } \operatorname{Re} s > 1,$$

where the outer product ranges over the primitive hyperbolic conjugacy classes in Γ of norm $p = NP > 1$. The infinite product converges absolutely; therefore it does not vanish in $\operatorname{Re} s > 1$. Differentiating it with respect to s gives

(10.39)
$$\frac{Z'}{Z}(s) = \sum_P \sum_{k=0}^{\infty} \frac{\log p}{p^{s+k} - 1}\,;$$

therefore

$$(2s - 1)^{-1}\frac{Z'}{Z}(s) - (2a - 1)^{-1}\frac{Z'}{Z}(a)$$

is exactly the contribution of the hyperbolic motions to the resolvent trace formula (10.35). This formula yields the analytic continuation of $(Z'/Z)(s)$ to the whole complex s-plane; the key point is that all poles of $(Z'/Z)(s)$ are simple and have integral residues. This is clear in every term of (10.35), except for the contributions from the elliptic classes and the identity motion, which have to be combined together into

$$\sum_m \left(m^{-1} \log m + (2s-1)^{-1} R_m(s) \right) - \psi(s) \left(\frac{|F|}{2\pi} + \sum_m \frac{1}{m} \right)$$

by an appeal to (10.33). It has already been observed after (10.34) that $R_m(s)$ has integral residues. The second part has poles at non-positive integers (see (10.18)) with residue

$$\frac{|F|}{2\pi} + \sum_m \frac{1}{m} = 2g - 2 + h + \ell \in \mathbb{Z}$$

by the Gauss-Bonnet formula (2.7), where g is the genus of $\Gamma \backslash \mathbb{H}$, h is the number of parabolic generators (cusps) and ℓ that of the elliptic ones. By virtue of the above properties we can define with no ambiguity a meromorphic function

$$F(s) = \exp \left(\int_a^s \frac{Z'}{Z}(u) \, du \right),$$

where the integration goes along any curve which joins a with s avoiding poles. Since $Z(s) = Z(a) F(s)$, this proves

Theorem 10.3. *The Selberg zeta-function $Z(s)$ defined for $\mathrm{Re}\, s > 1$ by (10.38) has a meromorphic continuation to the whole complex s-plane. In the half plane $\mathrm{Re}\, s \geq 1/2$ it is holomorphic and has zeros at the points s_j and $\overline{s_j}$ of order equal to the dimension of the λ_j-eigenspace except for $s = 1/2$, where $Z(s)$ has a zero or pole of order equal to twice the dimension of the $(1/4)$-eigenspace minus the number of inequivalent cusps.*

The remaining zeros and poles of $Z(s)$ in the half plane $\mathrm{Re}\, s < 1/2$ can be likewise determined by examining the resolvent trace formula (in order to interpret the continuous spectrum integral in (10.35), use the expansion (11.9) of $-\varphi'(s)/\varphi(s)$ into simple fractions and the residue theorem). Besides, one can also derive a functional equation of type

(10.40) $$Z(s) = \Psi(s) Z(1-s),$$

where $\Psi(s)$ is a certain meromorphic function of order 2 which can be written explicitly in terms of elementary functions, the Euler Γ-function and the Barnes G-function,

$$G(s+1) = (2\pi)^{s/2} e^{-s(s+1)/2 - \gamma s^2/2} \prod_{n=1}^{\infty} \left(1 + \frac{s}{n} \right)^n e^{-s + s^2/2n}.$$

As a matter of fact, one can attach to $Z(s)$ a finite number of local factors corresponding to the identity, the parabolic and the elliptic classes so that the complete zeta-function satisfies a simpler equation $Z^*(s) = Z^*(1 - s)$ (*cf.* [Vig1] and [Fi]).

If you will, the Selberg zeta-function satisfies an analogue of the Riemann hypothesis. However, the analogy with the Riemann zeta-function is superficial. First of all, the Selberg zeta function has no natural development into Dirichlet series. Furthermore, the functional equation (10.40) resists any decent interpretation as a kind of Poisson summation principle. Nevertheless, modern studies of $Z(s)$ have caused a lot of excitement in mathematical physics (see [Sa1]). At least, one may say that the dream of Hilbert and Pólya of connecting the zeros of a zeta-function with eigenvalues of a self-adjoint operator is a reality in the context of $Z(s)$.

10.9. Asymptotic law for the length of closed geodesics.

Perhaps the most appealing application of the Selberg trace formula is the evaluation of the length of closed geodesics in the Riemann surface $\Gamma\backslash\mathbb{H}$. Let us begin with a simple test function

$$g(x) = 2\left(\cosh\frac{x}{2}\right)e^{-2\delta\cosh x},$$

where $0 < \delta \leq 1$. Its Fourier transform is equal to

$$h(t) = 2\int_0^{+\infty}\left(\cosh(sx) + \cosh(1-s)x\right)e^{-2\delta\cosh x}\,dx$$
$$= 2\,K_s(2\delta) + 2\,K_{1-s}(2\delta) = \Gamma(s)\delta^{-s} + O(\delta^{-1/2}|\Gamma(s)|)\,,$$

if $1/2 \leq \operatorname{Re} s \leq 1$; clearly, $h(t)$ satisfies the conditions (1.63). All terms in the trace formula (10.37) contribute at most $O(\delta^{-1/2})$, except for the points of the discrete spectrum with $1/2 < s_j \leq 1$ and the primitive hyperbolic classes. A primitive hyperbolic class P of norm $p > 1$ contributes

$$\frac{p+1}{p-1}\,e^{-\delta\,(p+p^{-1})}\log p = \left(1 + O\left(\frac{1}{p}\right)\right)e^{-\delta p}\log p\,.$$

Estimating the error term trivially, we are left with the following result.

Theorem 10.4. *For any $\delta > 0$,*

(10.41) $$\sum_P e^{-\delta p}\log p = \sum_{1/2 < s_j \leq 1}\Gamma(s_j)\,\delta^{-s_j} + O(\delta^{-1/2})\,,$$

the implied constant depending on the group Γ.

Next, let us try another test function $g(x) = 2\,(\cosh(x/2))\,q(x)$, where $q(x)$ is even, smooth, supported on $|x| \leq \log(X + Y)$, and such that $0 \leq q(x) \leq 1$ and $q(x) = 1$ if $|x| \leq \log X$. The parameters $X \geq Y \geq 1$ will be chosen later. For $s = 1/2 + it$ in the segment $1/2 < s \leq 1$ we have

$$h(t) = \int_{-\infty}^{+\infty} \left(e^{sx} + e^{(1-s)x}\right) q(x)\,dx = s^{-1}X^s + O(Y + X^{1/2}),$$

and for s on the line $\operatorname{Re} s = 1/2$ we get by partial integration that

$$h(t) \ll |s|^{-1}X^{1/2}\min\{1, |s|^{-2}T^2\},$$

where $T = XY^{-1}$. Hence the discrete spectrum contributes

$$\sum_j h(t_j) = \sum_{1/2 < s_j \leq 1} s_j^{-1}X^{s_j} + O(Y + X^{1/2}T),$$

and the continuous spectrum contributes to the error term above. On the geometric side the identity motion contributes

$$\frac{|F|}{4\pi}\int_{-\infty}^{+\infty} h(t)\,\tanh(\pi t)\,t\,dt \ll X^{1/2}T.$$

The elliptic and parabolic classes contribute no more than the above bound. Gathering these estimates, we arrive at

$$(10.42) \qquad \sum_P q(\log p)\,\log p = \sum_{1/2 < s_j \leq 1} s_j^{-1}X^{s_j} + O(Y + X^{1/2}T).$$

We shall clean this formula up by exploiting the positivity of terms. First, subtracting (10.42) from that for $X + Y$ in place of X, we deduce that

$$\sum_{X < p < X+Y} \log p \ll Y + X^{1/2}T.$$

Hence, in (10.42) we can drop the excess over $p \leq X$ within the error term already present. Then we choose $Y = X^{3/4}$ to minimize the error term and obtain the following asymptotic expression:

Theorem 10.5 (Selberg). *For $X \geq 1$,*

$$(10.43) \qquad \sum_{p \leq X} \log p = \sum_{1/2 < s_j \leq 1} s_j^{-1}X^{s_j} + O(X^{3/4}),$$

where $p = NP$ denotes the norm of primitive hyperbolic classes, i.e. $\log p$ is the length of a closed geodesic in $\Gamma \backslash \mathbb{H}$ counted with multiplicity. The implied constant depends on the group.

An alternative approach to (10.43) makes use of analytic properties of the Selberg zeta-function $Z_\Gamma(s)$. Since $Z_\Gamma(s)$ satisfies the Riemann hypothesis (all zeros and poles other than $1/2 < s_j \leq 1$ are on the line $\operatorname{Re} s = 1/2$ or to the left), one should ask if (10.43) holds true with smaller error term. Here it is not clear if $O(x^{1/2+\varepsilon})$ is possible, as it is for the prime number theorem assuming the classical Riemann hypothesis. In the case of the modular group W. Luo and P. Sarnak [Lu-Sa] have recently established by refining the arguments of H. Iwaniec [Iw3] that

$$(10.44) \qquad \sum_{p \leq X} \log p = X + O(X^{7/10+\varepsilon}).$$

Any improvement upon the exponent $3/4$ is meaningful; it amounts to showing that there is a considerable regularity in the distribution of the zeros s_j on the critical line. In the proof of (10.44) Luo and Sarnak appeal to the Weil bound for Kloosterman sums (therefore, indirectly, to the Riemann hypothesis for curves) and to the recent estimate (8.46) of Hoffstein and Lockhart. The result has a connection with indefinite binary quadratic forms, since the lengths of closed geodesics on $SL_2(\mathbb{Z}) \backslash \mathbb{H}$ are given by $2 \log \varepsilon_d$ with multiplicity $h(d)$, where ε_d is the fundamental unit and $h(d)$ is the class number (*cf.* [Sa4]).

Chapter 11

The Distribution
of Eigenvalues

In this chapter we consider some fundamental issues of the Riemann surface $\Gamma\backslash\mathbb{H}$ concerning the eigenvalues of the Laplace operator. They are the subject of study in various fields of mathematics and physics. Hence, there are diverse techniques, taken from topology, differential geometry, partial differential equations, automorphic forms and number theory. Among these the Selberg trace formula is a conventional tool for establishing asymptotics in the spectrum. We shall apply the trace formula to prove Weyl's law, and then tackle the big problem of small eigenvalues.

11.1. Weyl's law.

The interplay between geometric invariants of the Riemann surface $\Gamma\backslash\mathbb{H}$ and the spectrum of the Laplacian is a wonderful gift of nature. Its content has been phrased by M. Kac [Kac] in the question: "Can one hear the shape of a drum?". M.-F. Vignéras [Vig2] has constructed strictly hyperbolic groups Γ_1, Γ_2 for which the surfaces $\Gamma_1\backslash\mathbb{H}$, $\Gamma_2\backslash\mathbb{H}$ are isospectral but not isometric (more precisely, they have the same spectrum with multiplicities, but the groups are not conjugate); thus the answer to the question is not always affirmative. Whatever the answer for a given group, it is easy to argue that reconstructing the surface out of the spectrum is not a very practical goal, since the set of eigenvalues is impossible to examine with the required precision. By the existing technology one can produce mainly statistical results. In practice, estimates and asymptotics involving the spectrum in suitable segments are quite useful.

First we wish to evaluate the counting function of the eigenvalues $\lambda_j = 1/4 + t_j^2$ in the discrete spectrum

$$N_\Gamma(T) = \#\{j : \ |t_j| \leq T\}\,.$$

By analogy, the continuous spectrum is measured by the integral

$$M_\Gamma(T) = \frac{1}{4\pi} \int_{-T}^{T} \frac{-\varphi'}{\varphi}\Big(\frac{1}{2} + it\Big)\, dt\,.$$

We shall shed more light upon $M_\Gamma(T)$ later. So far we have only shown that each part of the spectrum cannot be immensely large; specifically,

$$N_\Gamma(T) \ll T^2\,,$$
$$M_\Gamma(T) \ll T^2\,,$$

for any $T \geq 1$. These bounds are rather cheap by-products of Bessel's inequality (see (7.11), (10.13)), yet T^2 is the right order of magnitude. In order to get a lower bound or an asymptotic formula, we need the complete spectral decomposition of an automorphic kernel, which inevitably forces us to treat the quantities $N_\Gamma(T)$ and $M_\Gamma(T)$ together.

Let us apply the trace formula for the Fourier pair $h(t) = e^{-\delta t^2}$ and $g(x) = (4\pi\delta)^{-1/2}e^{-x^2/4\delta}$, where δ is a small positive parameter. The identity motion contributes

$$\frac{|F|}{4\pi} \int_{-\infty}^{+\infty} e^{-\delta t^2} \tanh(\pi t)\, t\, dt = \frac{|F|}{4\pi\delta} + O(1)\,.$$

The hyperbolic and the elliptic motions contribute a bounded quantity. The parabolic motions contribute

$$-(4\pi\delta)^{-1/2}h \log 2 - \frac{h}{2\pi} \int_{-\infty}^{+\infty} e^{-\delta t^2} \psi(1 + it)\, dt + \frac{1}{4} \operatorname{Tr}\Big(I - \Phi\big(\tfrac{1}{2}\big)\Big)\,.$$

By (B.11) we get

$$\psi(1 + it) + \psi(1 - it) = \log(1 + t^2) + O((1 + t^2)^{-1})\,;$$

hence the above integral is equal to (up to a bounded term)

$$2 \int_0^{+\infty} e^{-\delta t^2} \log t\, dt = \frac{1}{2\sqrt{\delta}} \int_0^{+\infty} e^{-t} \log\Big(\frac{t}{\delta}\Big)\, t^{-1/2}\, dt$$

$$= \frac{1}{2\sqrt{\delta}} \Big(\Gamma'\big(\tfrac{1}{2}\big) - \Gamma\big(\tfrac{1}{2}\big) \log \delta\Big)$$

$$= \frac{\sqrt{\pi}}{2\sqrt{\delta}} \big(-\gamma - \log 4\delta\big)\,.$$

From the above computations we conclude the following:

Theorem 11.1. *For any $\delta > 0$ we have*

$$
(11.1) \quad
\begin{aligned}
\sum_j e^{-\delta t_j^2} + \frac{1}{4\pi} \int_{-\infty}^{+\infty} \frac{-\varphi'}{\varphi}\left(\frac{1}{2} + it\right) e^{-\delta t^2}\, dt \\
= \frac{|F|}{4\pi\delta} + \frac{h \log \delta}{4\sqrt{\pi\delta}} - \frac{\gamma h}{4\sqrt{\pi\delta}} + O(1)\,,
\end{aligned}
$$

where $O(1)$ is bounded by a constant depending on the group.

By a Tauberian argument one deduces the following asymptotic (called Weyl's law):

Corollary 11.2. *As $T \to +\infty$ we have*

$$
(11.2) \qquad N_\Gamma(T) + M_\Gamma(T) \sim \frac{|F|}{4\pi} T^2\,.
$$

The strongest form of Weyl's law ever established for a general surface of curvature -1 is

$$
(11.3) \quad N_\Gamma(T) + M_\Gamma(T) = \frac{|F|}{4\pi} T^2 - \frac{h}{\pi} T \log T + c_\Gamma\, T + O(T(\log T)^{-1})\,,
$$

where c_Γ is a constant (*cf.* [He1, Theorem 2.28] and [Ve, Theorem 7.3]).

One sees from (11.2) that the total spectrum is quite large. As we cannot count separately the point and the continuous spectra, it remains an open question which one is larger (if either) in order of magnitude. The only answers we can get so far are for special groups by showing that $M_\Gamma(T)$ is much smaller than $N_\Gamma(T)$. For congruence groups Selberg has proved that

$$
(11.4) \qquad M_\Gamma(T) \ll T \log T,
$$

whence

$$
(11.5) \qquad N_\Gamma(T) = \frac{|F|}{4\pi} T^2 + O(T \log T).
$$

In this case the situation is clear. The determinant of the scattering matrix $\varphi(s)$ is shown to be a product of Dirichlet L-functions, so it is meromorphic of order 1, thus leading to (11.4). An immediate consequence of the resulting asymptotic (11.5) is that the congruence groups have infinitely many linearly independent cusp forms. This is a rather indirect argument. Sadly enough, not a single cusp form has yet been constructed for the modular group.

Nothing like (11.4) is known in general. Recent intensive studies, initiated by R. S. Phillips and P. Sarnak [Ph-Sa], then expanded by S. Wolpert [Wo], all indicate that the opposite situation is more likely to be true for generic groups, *i.e.* the cuspidal spectrum should be small. W. Luo [Lu] has supplied very concrete and sharp results to enhance these astonishing observations. Also the numerical computations by D. Hejhal [He2] tend to support such a theory. It would be interesting to isolate a finite volume quotient $\Gamma\backslash\mathbb{H}$ having a cusp but no cusp forms. The Phillips-Sarnak theory is still not complete. Had one showed that the multiplicities of eigenvalues, say $m(\lambda)$ — the dimension of eigenspaces, for the group $\Gamma_0(q)$ were bounded, the result in [Lu] would have reached the main objective of the theory. Unfortunately very little is known; by (11.3) one only infers that

$$(11.6) \qquad\qquad m(\lambda) \ll \frac{\lambda^{1/2}}{\log(\lambda + 3)} \, .$$

This easy bound is the best available today, even for the modular group, for which presumably $m(\lambda) = 1$. Any improvement upon (11.6) in the order of magnitude would be welcomed.

Before devoting greater attention to congruence groups, let us reveal what the quantity $M_\Gamma(T)$ really stands for in general. First we show some estimates for the scattering matrix $\Phi(s)$. For $\operatorname{Re} s > 1$ the entries are given by Dirichlet's series (3.21). Hence, the determinant is also given by a Dirichlet series (do not underestimate the significance of this fact!):

$$\varphi(s) = \left(\sqrt{\pi}\,\frac{\Gamma(s - 1/2)}{\Gamma(s)}\right)^h \sum_{n=1}^{\infty} a_n\, b_n^{-2s}$$

with $a_1 \neq 0$ and $0 < b_1 < b_2 < \cdots < b_n \to +\infty$. The series converges absolutely in $\operatorname{Re} s \geq 1 + \varepsilon$, so it is bounded, and therefore, by Stirling's formula,

$$b_1^{2s}\, \varphi(s) \ll |s|^{-h/2} \, .$$

Also notice that $\varphi(s)$ does not vanish in a half plane $\operatorname{Re} s > \sigma_0$ with σ_0 sufficiently large. In the half-plane $\operatorname{Re} s \geq 1/2$ it has a finite number of poles, all in the segment $1/2 < s_j \leq 1$. Following Selberg, we put

$$\varphi^*(s) = b_1^{2s-1}\varphi(s) \prod_{1/2 < s_j \leq 1} \frac{s - s_j}{s - 1 + s_j} \, .$$

The local factors are repeated with multiplicities of each s_j, so that $\varphi^*(s)$ is holomorphic in the half-plane $\mathrm{Re}\, s \geq 1/2$. We have

$$\varphi^*(s) \ll |s|^{-h/2} \qquad \text{in } \mathrm{Re}\, s \geq 1 + \varepsilon \,,$$
$$\varphi^*(s)\, \varphi^*(1-s) = 1 \qquad \text{in the whole } s\text{-plane} \,,$$
$$|\varphi^*(s)| = 1 \qquad \text{on the line } \mathrm{Re}\, s = 1/2 \,,$$
$$|\varphi^*(s)| \leq 1 \qquad \text{in the half plane } \mathrm{Re}\, s \geq 1/2 \,.$$

We appeal to Jensen's formula

$$\int_0^1 \log |f(e^{2\pi i\theta})|\, d\theta = \log |f(0)| - \sum_j \log |z_j|,$$

where $f(z)$ is any holomorphic function in $|z| < 1$ which is continuous in $|z| \leq 1$ with $f(0) \neq 1$, has no zeros on $|z| = 1$, and z_j ranges over the zeros in $|z| < 1$ repeated with multiplicity. By changing $z \mapsto (s-1)/s$ we map the unit disk $|z| \leq 1$ onto the half-plane $\mathrm{Re}\, s \geq 1/2$. Then, after removing a possible zero of $\varphi^*(s)$ at $s = 1$, we conclude by Jensen's formula for $f(z) = \varphi^*(1/(1-z))z^m$ that

$$-\sum_j{}' \log \left| \frac{z_j - 1}{z_j} \right| < +\infty,$$

where the summation ranges over the zeros of $\varphi^*(s)$ in $\mathrm{Re}\, s > 1/2$ different from 1. But a zero z_j corresponds to a pole $s_j = 1 - z_j$ by the functional equation; therefore, the above inequality can be written as

$$\sum_j{}' \log \left| \frac{1 - s_j}{s_j} \right| < +\infty,$$

where s_j ranges over the poles of $\varphi^*(s)$ in $\mathrm{Re}\, s < 1/2$ different from 0. Putting $s_j = \beta_j + i\gamma_j$ and applying

$$2 \log \left| \frac{1-s}{s} \right| = \log \left(1 + \frac{1 - 2\beta}{|s|^2} \right) = \frac{1 - 2\beta}{|s|^2} + O(|s|^{-2}),$$

we rewrite this inequality again as

(11.7) $$\sum_{\beta_j < 1/2}{}' (1 - 2\beta_j)\, |s_j|^{-2} < +\infty \,.$$

Now we can consider the Hadamard canonical product for $\varphi^*(s)$:

$$\varphi^*(s) = \prod_j \left(\frac{s - 1 + \bar{s}_j}{s - s_j} \right) e^{g(s)},$$

where $g(s)$ is a polynomial. Here a pole s_j is matched with the zero $1 - \bar{s}_j$, so the product converges by virtue of (11.7). Since $\varphi^*(s)$ is bounded in $\operatorname{Re} s \geq 1 + \varepsilon$, $g(s)$ is constant. Therefore

$$(11.8) \qquad -\frac{\varphi^{*\prime}}{\varphi^*}(s) = \sum_{\beta_j < 1/2} \left(\frac{1}{s - s_j} - \frac{1}{s - 1 + \bar{s}_j} \right).$$

Adding the poles in $1/2 < s_j \leq 1$, we conclude that

$$(11.9) \qquad -\frac{\varphi'}{\varphi}(s) = \sum_j \left(\frac{1}{s - s_j} - \frac{1}{s - 1 + \bar{s}_j} \right) + 2 \log b_1,$$

where s_j ranges over all poles of $\varphi(s)$ with proper multiplicity (notice that all poles of $\varphi(s)$ are in the strip $1 - \sigma_0 \leq \operatorname{Re} s \leq 1$).

On the critical line $s = 1/2 + it$ the terms of (11.8) are positive:

$$(1 - 2\beta_j)\left((1/2 - \beta_j)^2 + (t - \gamma_j)^2 \right)^{-1} > 0,$$

whence $-\varphi^{*\prime}(s)/\varphi^*(s) \geq 0$ and

$$(11.10) \qquad -\frac{\varphi'}{\varphi}(s) > 2 \log b_1 + O(|s|^{-2}).$$

Finally, by Cauchy's theorem, (11.9) reveals that the integral $M_\Gamma(T)$ is approximately equal to the number of poles of $\varphi(s)$ on the left of the critical line of height less than T up to an error term $O(T)$. If the number of such poles is of order of magnitude T^2, then most of them must concentrate along the critical line $\operatorname{Re} s = 1/2$ because of (11.7). Selberg [Se2] has established several stronger results of this kind. For the modular group, $\varphi(s)$ has poles at the complex zeros of $\zeta(2s)$; therefore, according to the Riemann hypothesis, these poles are on the line $\operatorname{Re} s = 1/4$.

11.2. The residual spectrum and the scattering matrix.

The poles of $\Phi(s)$ in the segment $1/2 < s_j \leq 1$ yield the so-called *residual eigenvalues* $\lambda_j = s_j(1 - s_j)$. There is always one at $s_0 = 1$, which corresponds to the lowest eigenvalue $\lambda_0 = 0$ with constant eigenfunction

$$u_0(z) = |F|^{-1/2}.$$

There are groups having many residual points arbitrarily close to 1, but they are not congruence groups.

Theorem 11.3. *The congruence groups have no residual spectrum besides the obvious point* $s_0 = 1$.

This result can be proved in a number of ways. Clearly it is enough to consider the principal congruence group $\Gamma(N)$. In this case one can compute the scattering matrix $\Phi(s)$ explicitly in terms of the Riemann zeta-function and Dirichlet L-series, and from this the claim is deduced. A complete computation is quite involved. However, since all poles appear on the diagonal of $\Phi(s)$ (see Theorem 6.10), one can save work by computing only the diagonal entries

$$\varphi_{\mathfrak{a}\mathfrak{a}}(s) = \sqrt{\pi}\,\frac{\Gamma(s-1/2)}{\Gamma(s)} \sum_c c^{-2s}\, \mathcal{S}_{\mathfrak{a}\mathfrak{a}}(0,0;c)\,.$$

For the cusp $\mathfrak{a} = \infty$ it gives

$$\varphi_{\infty\infty}(s) = \sqrt{\pi}\,\frac{\Gamma(s-1/2)}{\Gamma(s)}\, N^{-1} \sum_{c \equiv 0\,(\mathrm{mod}\,N^2)} \varphi(c)\, c^{-2s}$$

$$= \sqrt{\pi}\,\frac{\Gamma(s-1/2)}{\Gamma(s)}\,\frac{\zeta(2s-1)}{\zeta(2s)}\,\frac{\varphi(N)}{N^{4s}} \prod_{p\mid N} \left(1 - \frac{1}{p^{2s}}\right)^{-1};$$

hence $\varphi_{\infty\infty}(s)$ has no poles in $\operatorname{Re} s \geq 1/2$ except for a simple one at $s = 1$. The same can be shown for other cusps.

Another approach proceeds directly from the Eisenstein series

$$E_{\mathfrak{a}}(\sigma_{\mathfrak{a}} z, s) = \sum_{\tau \in \Gamma_\infty \backslash \sigma_{\mathfrak{a}}^{-1} \Gamma_{\mathfrak{a}}} (\operatorname{Im} \tau z)^s\,.$$

Using a parametrization of the above cosets (one does not need to be very explicit), the series splits into a finite number of Epstein zeta-functions

$$\sum_{m,n} Q(m,n)^{-s},$$

where Q is a positive definite quadratic form with rational coefficients and m, n range over co-prime integers in an arithmetic progression. Every such series has an analytic continuation to $\operatorname{Re} s > 1/2$ by Poisson's summation with at most a simple pole at $s = 1$.

The scattering matrix $\Phi(s)$ has been computed completely in some cases. For instance, if $\Gamma = \Gamma_0(p)$ and p is prime, we derive easily, exploiting the computations following Theorem 2.7, that

$$(11.11) \qquad \Phi(s) = \begin{pmatrix} \varphi_{\infty\infty}(s) & \varphi_{\infty 0}(s) \\ \varphi_{0\infty}(s) & \varphi_{00}(s) \end{pmatrix} = \varphi(s)\, N_p(s),$$

where

$$N_p(s) = (p^{2s} - 1)^{-1} \begin{pmatrix} p - 1 & p^s - p^{1-s} \\ p^s - p^{1-s} & p - 1 \end{pmatrix}$$

and $\varphi(s)$ is the scattering matrix for the modular group given by (3.24). This can be written in the symmetric fashion (see (3.27) and (3.28))

$$\Phi(s) = M(s)^{-1} M(1 - s),$$

where

$$M(s) = \pi^{-s} \, \Gamma(s) \, \zeta(2s) \begin{pmatrix} 1 & p^s \\ p^s & 1 \end{pmatrix}.$$

The above results are consistent with these due to D. Hejhal [He1] (see also [Hu1]). For the group $\Gamma_0(N)$ with N squarefree, Hejhal provides us with

(11.12) $$\Phi(s) = \varphi(s) \bigotimes_{p|N} N_p(s).$$

11.3. Small eigenvalues.

How small can the first positive eigenvalue λ_1 in the discrete spectrum of $\Gamma \backslash \mathbb{H}$ possibly be? Undoubtedly it is an important question, and an intricate one too. In particular, we wish to know if $\lambda_j \geq 1/4$, which would mean that the positive discrete spectrum lies on the continuous one (if Γ has cusps). We shall call λ_j with $0 < \lambda_j < 1/4$ *exceptional;* equivalently $\lambda_j = s_j(1 - s_j)$ with $1/2 < s_j < 1$, emphasising that they are not welcomed. In practice the exceptional eigenvalues distort rather than simplify results. That exceptional eigenvalues exist, both cuspidal and residual ones, for some groups was known to A. Selberg, proved by B. Randol and constructed by M. N. Huxley (just to name a few from a long list of investigators).

A great deal of research concerns compact, smooth Riemann surfaces. Suppose F is one of those having curvature -1. By the uniformization theorem, F can be represented as a quotient $\Gamma \backslash \mathbb{H}$ for some hyperbolic group of signature $(g; 0; 0)$ with $g \geq 2$. In this case the Gauss-Bonnet theorem asserts that $|F| = 4\pi(g - 1)$, and Weyl's law implies $\lambda_j \to 4\pi |F|^{-1} j$ as $j \to \infty$. R. Schoen, S. Wolpert and S. T. Yau showed that λ_{2g-3} can be as small as one likes. In the other direction, P. Buser proved that λ_{4g-2} is never exceptional, *i.e.* the lower bound

(11.13) $$\lambda_{4g-2} \geq \frac{1}{4}$$

always holds true, while λ_{4g-3} can be exceptional for arbitrary $g \geq 2$.

It is remarkable that the first eigenvalue is bounded above by an absolute constant; for example, we obtain (essentially by variational calculus)

$$(11.14) \qquad \lambda_1 \le 2\,\frac{g+1}{g-1} \le 6\,,$$

which is due to P. C. Yang and S. T. Yau. It is also possible to estimate λ_1 in terms of the diameter d of F alone:

$$\left(4\pi\,\sinh\frac{d}{2}\right)^{-2} < \lambda_1 < \frac{1}{4} + \frac{4\pi^2}{d^2}\,.$$

These estimates are due to P. Buser and S.-Y. Cheng, respectively.

Some of the methods used to establish the above results adapt well to general quotients $F = \Gamma\backslash\mathbb{H}$ of finite volume. Thus P. G. Zograf [Zo] has generalized (11.14) to show that

$$(11.15) \qquad \lambda_1 \le 8\pi\,(g+1)\,|F|^{-1}$$

for any F which is not compact and has volume $|F| \ge 32\pi\,(g+1)$; in particular, such a surface has exceptional eigenvalues. Suppose the corresponding group Γ is a subgroup of $PSL_2(\mathbb{Z})$ of finite index $n \ge 1$ and that $F = \Gamma\backslash\mathbb{H}$ has exactly one cusp (such groups have been studied by H. Petersson [Pe2]). One shows that the Eisenstein series for Γ and $PSL_2(\mathbb{Z})$ differ by the factor n^{-s}; therefore Γ has only $\lambda_0 = 0$ in the residual spectrum. Consequently, λ_1 is cuspidal, and, using (11.15), λ_1 can be made very small; just take a subgroup of signature $(0; 2, \ldots, 2; 1)$ with a large number of elliptic classes of order 2.

For congruence groups it is a quite different story.

Conjecture (A. Selberg [Se1]). *There is no exceptional spectrum for congruence groups, i.e. one has the bound*

$$(11.16) \qquad \lambda_1 \ge \frac{1}{4}\,.$$

The conjecture is known to be true for groups of small level (see M. N. Huxley [Hu2]). In the case of the modular group H. Maass [Ma] and W. Roelcke [Ro] have established (by slightly different methods) somewhat sharper bounds. We apply Roelcke's argument to prove the following fact.

Theorem 11.4. *Let Γ be a subgroup of finite index in the modular group. Denote by q the maximal width of cuspidal zones. Then the cuspidal eigenvalues satisfy the lower bound*

$$(11.17) \qquad \lambda \ge \frac{3}{2}\left(\frac{\pi}{q}\right)^2\,.$$

Proof. Let u be a cusp form for Γ with $\|u\| = 1$. By Lemma 4.1 (see (4.3)) the eigenvalue of u is equal to the energy integral (Dirichlet)

$$\lambda = \int_F |y\,\nabla u(z)|^2\,d\mu z,$$

where F is any fundamental domain of Γ. Choose $F = \bigcup \gamma_\nu F'$, where F' is the standard fundamental polygon for the modular group

$$F' = \left\{ z : \ |x| < \frac{1}{2},\ |z| > 1 \right\}$$

and γ_ν ranges over the coset representatives. Consider also another fundamental polygon for the modular group obtained from F' by applying the involution $\omega = \begin{pmatrix} & -1 \\ 1 & \end{pmatrix}$. Observe that $F'' = F' \cup \omega F'$ covers the strip $\{z : \ |x| < 1/2,\ y > \sqrt{3}/2\}$, which contains F'. Take the unions $F_1 = \bigcup (F' + b)$ and $F_2 = \bigcup (F'' + b)$ over $0 \le b < B$. We obtain

$$2\lambda B \ge \sum_\nu \int_{\gamma_\nu F_2} |y\,\nabla u(z)|^2\,d\mu z$$

$$= \sum_\nu \int_{F_2} |(y\,\nabla)u(\gamma_\nu z)|^2\,d\mu z$$

$$\ge \sum_\nu \int_{\sqrt{3}/2}^{+\infty} \int_0^B |(y\,\nabla)u(\gamma_\nu z)|^2\,d\mu z\,.$$

Here for each ν we apply the Fourier expansion of $u(\gamma_\nu z)$ in the cusp $\mathfrak{a}_\nu = \gamma_\nu \infty$,

$$u(\gamma_\nu z) = \sum_{n \ne 0} c_{\nu n}(y)\,e\!\left(\frac{nx}{q_\nu}\right),$$

where q_ν is the width. We have

$$\nabla u(\gamma_\nu z) = \left[\sum_{n \ne 0} \frac{2\pi n}{q_\nu} c_{\nu n}(y)\,e\!\left(\frac{nx}{q_\nu}\right),\ \sum_{n \ne 0} c'_{\nu n}(y)\,e\!\left(\frac{nx}{q_\nu}\right) \right].$$

Choose B divisible by all q_ν. We obtain

$$\int_{\sqrt{3}/2}^{+\infty} \int_0^B |(y\,\nabla)u(\gamma_\nu z)|^2\,d\mu z \ge B \int_{\sqrt{3}/2}^{+\infty} \sum_{n \ne 0} \left| \frac{2\pi n}{q_\nu} c_{\nu n}(y) \right|^2 dy$$

$$\ge 3\left(\frac{\pi}{q}\right)^2 B \int_{\sqrt{3}/2}^{+\infty} \sum_{n \ne 0} |c_{\nu n}(y)|^2 y^{-2}\,dy$$

$$= 3\left(\frac{\pi}{q}\right)^2 \int_{\sqrt{3}/2}^{+\infty} \int_0^B |u(\gamma_\nu z)|^2\,d\mu z$$

$$\ge 3\left(\frac{\pi}{q}\right)^2 \int_{\gamma_\nu F_1} |u(z)|^2\,d\mu z\,.$$

Summing over the coset representatives γ_ν, we get $2\lambda B \geq 3\pi^2 q^{-2} B$, whence (11.17).

For the modular group Theorem 11.4 yields $\lambda_1 \geq 3\pi^2/2 = 14.80\ldots$, but the true value is $\lambda_1 = 91.14\ldots$ according to numerical computations of D. Hejhal (see [He1, Appendix C]). If the cuspidal widths are ≤ 7, we get from (11.7) that $\lambda_1 > 3/11$. In particular, this proves

Corollary 11.5. *There is no exceptional spectrum for congruence groups of level ≤ 7.*

The Selberg eigenvalue conjecture is also true for some non-congru- ence groups. P. Sarnak [Sa3] has proved it for all Hecke triangle groups, using the Courant nodal line technique (see also M. N. Huxley [Hu2]).

The Roelcke technique is wasteful. It cannot work for groups of very large level, since there are a lot of eigenvalues near to $1/4$. Indeed, by employing the trace formula for the group $\Gamma_0(q)$ with a suitable test function we can show that

$$(11.18) \qquad \#\left\{ j : \frac{1}{4} \leq \lambda_j < \frac{1}{4} + c \, (\log q)^{-2} \right\} \asymp |F| \, (\log q)^{-3},$$

where c is a large constant. Hence it follows that the dimension of the eigenspace $\lambda = 1/4$ satisfies $m(1/4) \ll |F|(\log|F|)^{-3}$, so it is quite a bit smaller than the volume.

A. Selberg [Se1] has established a remarkable lower bound (by an absolute positive constant).

Theorem 11.6. *For any congruence group we have*

$$(11.19) \qquad\qquad \lambda_1 \geq \frac{3}{16} \, .$$

Proof. Apply Theorem 9.2 and Weil's bound (2.25) for Kloosterman sums. This shows that the series $Z_s(m,n)$ converges absolutely in $\operatorname{Re} s > 3/4$, so its poles at s_j have $\operatorname{Re} s_j \leq 3/4$; hence, $\lambda_j = s_j(1 - s_j) \geq 3/16$.

A better bound was established in the summer of 1994 by W. Luo, Z. Rudnick and P. Sarnak [Lu-Ru-Sa]; $\lambda_1 \geq 21/100$, by using properties of the Rankin-Selberg convolution L-functions on GL_3. We can obtain by arguments entirely within the GL_2 theory the slightly weaker bound $\lambda_1 \geq 10/49$. The best result so far established is

$$\lambda_1 \geq \frac{1}{4} - \left(\frac{7}{64} \right)^2,$$

due to H. Kim and P. Sarnak [Ki-Sa]. Besides high rank groups, in all these improvements one also exploits the root number of relevant L-functions and estimates for hyper Kloosterman sums; therefore the Riemann hypothesis for varieties supplies the power indirectly.

11.4. Density theorems.

In practice a few exceptional eigenvalues do not cause a problem, but a larger number of them may ruin results. Therefore it is important to investigate the distribution of the points s_j in the segment $1/2 < s_j < 1$ from a statistical point of view. A natural way would be to count all points s_j with certain weights such that the larger s_j is, the heavier the weight that is attached to s_j. It turns out that an inequality of the form

$$(11.20) \qquad \sum_{1/2 < s_j < 1} |F|^{c(s_j - 1/2)} \ll |F|^{1+\varepsilon}$$

is most appropriate for applications. We shall call any such result with exponent $c > 0$ a *density theorem*. A density theorem with sufficiently large exponent completely substitutes for the Selberg eigenvalue conjecture, in the same fashion as the density theorems for zeros of L-functions serve in the absence of the Riemann hypothesis.

Let us first deal with the simplest case of a hyperbolic group. Our approach is an excercise with the trace formula (10.37). Take the test function

$$(11.21) \qquad h(t) = \left(\frac{\sin tL}{tL} \right)^4$$

so that the Fourier transform $g(x)$ is supported in the segment $[-4L, 4L]$. Choose $4L$ to be the length of the shortest closed geodesic on $\Gamma \backslash \mathbb{H}$, or any smaller number. Then there is no contribution from the hyperbolic classes to the trace formula (the parabolic and elliptic classes do not exist by assumption), and therefore we are left with the cute identity

$$\sum_j \left(\frac{\sin t_j L}{t_j L} \right)^4 = \frac{|F|}{2\pi} \int_0^{+\infty} \left(\frac{\sin tL}{tL} \right)^4 \tanh(\pi t) \, t \, dt \, .$$

On the right side apply the bound $\tanh(\pi t) \le \pi t$, getting $\pi |F| (2L)^{-3}$. On the left side discard all but the exceptional points, and for them use the bound $\sinh x \ge x(2x+1)^{-1} e^x$ to get

$$\frac{\sin t_j L}{t_j L} \ge \frac{e^{|t_j| L}}{2|t_j| L + 1} \ge \frac{e^{|t_j| L}}{L + 1} \, .$$

Inserting these estimates into the above identity, we establish the following density theorem:

$$(11.22) \qquad \sum_{1/2 < s_j \leq 1} e^{4(s_j - 1/2)L} \leq \pi |F| (2L)^{-3} (L+1)^4.$$

Since the point $s_0 = 1$ is always present, $e^{2L} \leq \pi |F| (2L)^{-3} (L+1)^4$. Hence we infer the following bound for the length of the shortest closed geodesic: $L \leq \frac{1}{2} \log(|F| \log |F|)$. Recall that $|F| = 4\pi(g-1)$ by the Gauss-Bonnet formula.

Next we establish a much deeper density theorem for the Hecke congruence group $\Gamma_0(q)$. It is relatively easy to derive (11.20) with exponent $c = 2$ from the Selberg trace formula (10.37). However, our goal is $c = 4$. This time it will be an exercise with the Kloosterman sums formula (9.12). Let us use the same test function (11.21) with $L \geq 4$. From (9.9), arguing as above we get $h_0 \leq \pi L^{-3}$. To estimate the transform $h^+(x)$ we use the contour integral

$$h^+(x) = -i \int_{(\sigma)} J_{2s-1}(x) \, h\big(i(s - \tfrac{1}{2})\big) \frac{2s-1}{\sin \pi s} \, ds$$

with $1/2 \leq \sigma < 1$. By the trivial estimate $|J_\nu(x)| \leq \sqrt{\pi} |x^\nu \Gamma(\nu+1/2)^{-1}|$ this yields $h^+(x) \ll (x e^{2L})^{2\sigma - 1}$. Furthermore, by Weil's bound for Kloosterman sums (2.25) we get

$$\sum_{c \equiv 0 \,(\mathrm{mod}\ q)} c^{-2\sigma} |\mathcal{S}(n, m; c)| \ll \tau(nq) (n, q)^{1/2} \, q^{-2\sigma + 1/2} \left(\sigma - \frac{3}{4}\right)^{-2}$$

uniformly in $3/4 < \sigma < 1$. Hence, choosing $\sigma = 3/4 + (L + \log 2nq)^{-1}$, we obtain

$$(11.23) \qquad \sum_{j>0} \left(\frac{\sin t_j L}{t_j L}\right)^4 |\nu_j(n)|^2$$

$$\ll L^{-3} + e^L n^{1/2} (n, q)^{1/2} q^{-1} \tau(nq) (\log 2nq)^2$$

(the continuous spectrum integrals are dropped due to the positivity).

Fixing L in (11.23) and dropping all but one term, we infer

Corollary 11.6. *The Fourier coefficient $\nu_j(n)$ of a normalized cusp form u_j for the group $\Gamma_0(q)$ satisfies the bound*

(11.24) $$\nu_j(n) \ll \lambda_j\, n^{1/4}\, \tau(n)\, \log 2n,$$

where the implied constant is absolute.

Choosing $n = 1$ and $L = 4 + \log q$ in (11.23), we obtain another inequality:

(11.25) $$\sum_{1/2<s_j<1} q^{4(s_j-1/2)}|\nu_j(1)|^2 \ll \tau(q)\,\log^6 q\,.$$

Hence, the density theorem (11.2) with exponent $c = 4$ would follow if for the first coefficient we had the lower bound

(11.26) $$|\nu_j(1)|^2 \gg q^{-1-\varepsilon}.$$

However, such a bound is false for some individual j. Clearly, one can make $\nu_j(1)$ vanish if the λ_j-eigenspace has dimension > 1. Therefore, for (11.26) to be true, a basis of cusp forms must be carefully selected. To this end, choose the Hecke basis as described in Section 8.5. Every form in the basis is of type $v(dz)$, where $v(z)$ is a newform on $\Gamma_0(r)$ with $dr|q$ (all forms are normalized with respect to the inner product in $\mathcal{L}(\Gamma_0(q)\backslash\mathbb{H})$, regardless of the level). By our lower bound (8.43) the first Fourier coefficient of $v(z)$ satisfies (11.26) (take into account our normalization). The other forms $v(dz)$ of divisor $d \neq 1$ do not contribute to (11.25), yet they have a common Laplace eigenvalue already accounted for from $v(z)$. Since the missing factor of multiplicity does not exceed $\tau(q) \ll q^\varepsilon$, (11.25) and (11.26) for newforms yield

Theorem 11.7 (Density Theorem). *The exceptional eigenvalues for $\Gamma_0(q)$ satisfy*

(11.27) $$\sum_{1/2<s_j<1} q^{4(s_j-1/2)} \ll q^{1+\varepsilon}.$$

Equivalently, for any $\sigma \geq 1/2$ we have

(11.28) $$\#\{j > 0:\ s_j > \sigma\} \ll q^{3-4\sigma+\varepsilon}.$$

(This is my favorite estimate for the cardinality of the empty set!)

Chapter 12

Hyperbolic Lattice-Point Problems

The lattice-point problem is described in general as follows. Let X be a topological space acted on discontinuously by a group Γ. Let D be a domain in X and let z be a point in X. The problem is to estimate the number of points of the orbit $\Gamma z = \{\gamma z : \gamma \in \Gamma\}$ which meet D. For a sufficiently regular domain D one expects that this number approximates the area of D with respect to a certain measure on X (the Haar measure if X is the homogeneous space of a Lie group).

A classical example is the euclidean plane $X = \mathbb{R}^2$ acted on by the group $\Gamma = \mathbb{Z}^2$ of integral translations. In this case we ask how many integer points are in D. The elementary method of packing with the unit square works fine. If D is a disc of radius $r = x^{1/2}$ centered at the origin, then one deals with the *Gauss circle problem*, and the packing method leads to

$$(12.1) \qquad \sum_{m \le x} r(m) = \pi x + O(x^{1/2}),$$

where $r(m)$ denotes the number of ways that m can be represented as the sum of two squares

$$(12.2) \qquad r(m) = \#\{a, b \in \mathbb{Z} : a^2 + b^2 = m\}.$$

If D is the area under a hyperbola, the problem is known as the *Dirichlet divisor problem*. We obtain

$$(12.3) \qquad \sum_{m \le x} \tau(m) = x \log x + (2\gamma - 1)\, x + O(x^{1/2}),$$

where $\tau(m)$ denotes the number of positive divisors of m,

$$(12.4) \qquad \tau(m) = \#\{a, b \in \mathbb{N} :\ ab = m\}.$$

In both asymptotics the error term has been sharpened to $O(x^{\theta+\varepsilon})$ with various θ many times by methods of exponential sums. The best exponent should be $\theta = 1/4$. M. N. Huxley [Hu3] got $\theta = 23/73$.

In this chapter we consider the lattice-point problem in the hyperbolic plane $\mathbb{H}$. In view of the isoperimetric inequality (1.14) the packing method is insufficient, because the area of a regular domain is comparable with the length of the boundary (the negative curvature causes the problem). The spectral theorem will show a lot more power here.

The lattice-point problem on $\mathbb{H}$ with respect to a group Γ is nothing other than a case of evaluating the automorphic kernel $K(z, w)$ for a properly chosen function $k(u)$. Recall the spectral expansion (7.17). The points s_j of the discrete spectrum in the segment $1/2 < s_j \le 1$ yield the main term. The remaining points of the discrete and continuous spectra lie on the critical line $\operatorname{Re} s = 1/2$. The contribution of these spectra is estimated by means of (7.10) using Cauchy's inequality. We obtain in general that

$$(12.5) \quad K(z, w) = \sum_{1/2 < s_j \le 1} h(t_j)\, u_j(z)\, \bar{u}_j(w) + O\Big(\int_0^{+\infty} (t+1)\, H(t)\, dt \Big),$$

where $H(t)$ is any decreasing majorant of $|h(t)|$, and the implied constant depends on the group Γ and the points z, w. It remains to choose the function $k(u)$ and survey its transform $h(t)$.

For the hyperbolic circle problem the aim is to estimate the number

$$P(X) = \#\Big\{\gamma \in \Gamma :\ 4u(\gamma z, w) + 2 \le X\Big\}.$$

Recall that $4u + 2 = e^\rho + e^{-\rho}$, where ρ is the hyperbolic distance. In this case (12.5) will yield

Theorem 12.1. *Let Γ be a finite volume group. For $X \ge 2$, we have*

$$(12.6) \quad P(X) = \sum_{1/2 < s_j \le 1} 2\pi^{1/2}\, \frac{\Gamma(s_j - 1/2)}{\Gamma(s_j + 1)}\, u_j(z)\, \bar{u}_j(w)\, X^{s_j} + O(X^{2/3}).$$

Note that the dominant term in (12.6) is attained at $s_0 = 1$, which corresponds to the lowest eigenvalue $\lambda_0 = 0$ with constant eigenfunction $u_0(z) = u_0(w) = |F|^{-1/2}$, so it contributes $2\pi|F|^{-1}X$.

Proof. Naturally, one would like to take $k(u) = 1$ if $u \le (X-2)/4$ and $k(u) = 0$ elsewhere, but this kernel does not yield strong results. Rather, we take $k(u)$ whose graph is

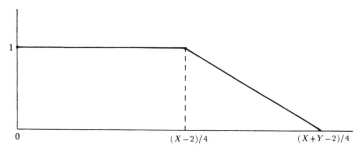

Figure 11. The test function $k(u)$.

with Y to be chosen later subject to $X \ge 2Y \ge 2$. This gives only a majorization

$$(12.7) \qquad P(X) \le 2\,K(z,w)\,.$$

We let the reader prove that the Selberg/Harish-Chandra transform of $k(u)$ satisfies

$$(12.8) \qquad h(t) = \pi^{1/2}\frac{\Gamma(s-1/2)}{\Gamma(s+1)}\,X^s + O(Y + X^{1/2})$$

if $1/2 < s \le 1$, where the implied constant depends on s, and

$$(12.9) \qquad h(t) \ll |s|^{-5/2}\left(\min\{|s|,T\} + \log X\right)X^{1/2}$$

if $\operatorname{Re} s = 1/2$, where $T = XY^{-1}$ and the implied constant is absolute.

Given these evaluations, we get by (12.5) that

$$K(z,w) = \sum_{1/2 < s_j \le 1} \pi^{1/2}\frac{\Gamma(s_j - 1/2)}{\Gamma(s_j + 1)}\,u_j(z)\,\bar{u}_j(w)\,X^{s_j} + O(Y + XY^{-1/2})\,.$$

This yields an upper bound for $P(X)$ through (12.7). A lower bound of the same type is obtained by applying the above result with X replaced by $X - Y$. Now clearly, the optimal choice is $Y = X^{2/3}$, giving (12.6).

For special points and groups the hyperbolic lattice-point problem can be stated in an arithmetic fashion. Let us take $z = w = i$. Then for $\gamma = \begin{pmatrix} a & b \\ c & d \end{pmatrix}$ we have $4\,u(\gamma i, i) + 2 = a^2 + b^2 + c^2 + d^2$. Therefore, if Γ is the modular group, then $P(X)$ counts the integer points (a, b, c, d) on the hypersurface

$$(12.10) \qquad ad - bc = 1$$

within the ball

$$(12.11) \qquad a^2 + b^2 + c^2 + d^2 \le X\,.$$

In this case Theorem 12.1 together with Corollary 11.5 gives

Corollary 12.2. *If* Γ *is the modular group, then*

(12.12) $$P(X) = 6X + O(X^{2/3}).$$

Another interesting case is for the group

(12.13) $$\Gamma = \left\{ \begin{pmatrix} a & b \\ c & d \end{pmatrix} \in SL_2(\mathbb{Z}) : \begin{array}{l} a \equiv d \ (\mathrm{mod}\ 2) \\ b \equiv c \ (\mathrm{mod}\ 2) \end{array} \right\}.$$

This is a subgroup of index 3 in the modular group; in fact, Γ is conjugate to $\Gamma_0(2)$, namely $\Gamma = \left(\begin{smallmatrix} 1 & -1 \\ & 1 \end{smallmatrix}\right)^{-1} \Gamma_0(2) \left(\begin{smallmatrix} 1 & -1 \\ & 1 \end{smallmatrix}\right)$. There are no exceptional points s_j by Corollary 11.5, so Theorem 12.1 gives

(12.14) $$P(X) = 2X + O(X^{2/3}).$$

In this case the linear map $a + d = 2k$, $a - d = 2\ell$, $b + c = 2m$, $b - c = 2n$ yields integers k, ℓ, m, n, and it transforms (12.10) into $k^2 - \ell^2 - m^2 + n^2 = 1$ and (12.11) into $k^2 + \ell^2 + m^2 + n^2 \le X/2$. Hence, it follows that

$$P(4x + 2) = \sum_{m \le x} r(m)\, r(m + 1).$$

Combining this with (12.14), we obtain (compare this with results in Chapter 0)

Theorem 12.3. *For* $x \ge 1$ *we have*

(12.15) $$\sum_{m \le x} r(m)\, r(m + 1) = 8x + O(x^{2/3}).$$

For the modular group the lattice-point problem can be generalized to count integer points on the hypersurface

(12.16) $$ad - bc = n,$$

where n is a fixed positive integer. Denote by $P_n(x)$ the number of such points within the ball (12.11).

Theorem 12.4. *For* $X \ge n \ge 1$ *we have*

(12.17) $$P_n(X) = 6\left(\sum_{d \mid n} d^{-1} \right)\left(X + O(n^{1/3} X^{2/3}) \right),$$

where the implied constant is absolute.

For the proof we need the Hecke operator $T_n : \mathcal{A}(\Gamma \backslash \mathbb{H}) \to \mathcal{A}(\Gamma \backslash \mathbb{H})$ defined by (see Section 8.5)

(12.18) $$(T_n f)(z) = \frac{1}{\sqrt{n}} \sum_{\tau \in \Gamma \backslash \Gamma_n} f(\tau z),$$

where Γ is the modular group and

$$\Gamma_n = \left\{ \begin{pmatrix} a & b \\ c & d \end{pmatrix} : a, b, c, d \in \mathbb{Z}, \ ad - bc = n \right\} .$$

Observe that $P_n(X)$ counts points of the orbit $\Gamma_n(z) = \{\gamma z : \gamma \in \Gamma_n\}$ in a disc. Indeed, for $\gamma \in \Gamma_n$ we have

$$4\,u(\gamma i, i) + 2 = \frac{1}{n}\,(a^2 + b^2 + c^2 + d^2) \leq \frac{X}{n} .$$

Therefore, in general the problem boils down to estimating the kernel

$$K_n(z, w) = \sum_{\gamma \in \Gamma_n} k(\gamma z, w) .$$

By (12.18) it follows that $K_n(z, w) = \sqrt{n}\, T_n\, K(z, w)$. On the other hand, applying T_n to the spectral decomposition (7.17), we infer that (asumming that the u_j are eigenfunctions of T_n, see Section 8.6)

$$n^{-1/2} K_n(z, w) = \sum_j \lambda_j(n)\, h(t_j)\, u_j(z)\, \bar{u}_j(w)$$

$$+ \frac{1}{4\pi} \int_{-\infty}^{+\infty} \eta_t(n)\, h(t)\, E\!\left(z, \frac{1}{2} + it\right) \overline{E}\!\left(w, \frac{1}{2} + it\right) dt .$$

Now insert the trivial bound $|\lambda_j(n)| \leq \lambda_0(n) = \sigma(n)\, n^{-1/2}$. The rest of the proof proceeds as in the case $n = 1$, but with X replaced by X/n. It gives

$$P_n(X) = 6\,\lambda_0(n)\, \sqrt{n} \left(\frac{X}{n} + O\!\left(\left(\frac{X}{n} \right)^{2/3} \right) \right)$$

thus completing the proof of Theorem 12.4.

If one applies all these results to the group (12.13) and transforms the points as in the proof of Theorem 12.3, then one ends up with

Theorem 12.5. *If n is odd and $x \geq n \geq 1$, then*

$$(12.19) \qquad \sum_{m \leq x} r(m)\, r(m + n) = 8 \left(\sum_{d \mid n} d^{-1} \right) \left(x + O(n^{1/3} x^{2/3}) \right) .$$

Remark. The error term $O(X^{2/3})$ in the above results has never been improved even slightly for any group. Recently F. Chamizo [Ch] has established numerous sharper results on average, while R. Phillips and Z. Rudnick [Ph-Ru] gave insightful analysis of the limit for various error terms relevant to the hyperbolic circle problem. Both works suggest that the exponent $2/3$ might be lowered to any number greater than $1/2$.

Chapter 13

Spectral Bounds
for Cusp Forms

13.1. Introduction.

The eigenfunctions of the Laplace operator on a riemannian manifold are
of great interest for theoretical physicists working in quantum mechanics.
The square-integrable eigenstates are particularly meaningful. How do they
behave on high energy levels—that is, in the limit with respect to the eigenvalues? Do they concentrate onto specific submanifolds, or sets such as
closed geodesics when being on distinguished energy levels, and if so, what
is the distribution law for these levels? For physicists, if the individual
eigenstates behave like random waves, this is a manifestation of quantum
chaos.

A simpler question, yet not easy to answer, is how large the eigenstates
can possibly be in terms of the spectrum. The case of the torus $\mathbb{Z}^2\backslash\mathbb{R}^2$
shows that all eigenstates of the standard basis are uniformly bounded.
But this is not true on other manifolds such as the sphere S^2, on which
the eigenfunctions given by the Legendre polynomials (spherical harmonics)
take relatively large values at special points (see the concluding remarks
in Appendix B.3). However, this phenomenon seems to be much weaker if
the manifold is negatively curved, such as the quotient space $\Gamma\backslash\mathbb{H}$ of the
hyperbolic plane modulo a finite volume group.

In this chapter (joint work with P. Sarnak [Iw-Sa]) we demonstrate how
to break the standard upper bound for cusp forms u_j on the modular group.
The standard bound for u_j is $O(\lambda_j^{1/4})$, which follows easily from the spectral
decomposition for a well-chosen automorphic kernel. One can also infer the

standard bound in a number of other ways and for general groups (see (8.3')). In order to strengthen this bound in the case of the modular group we exploit a bit of arithmetic through the Hecke correspondence. A conjecture we make about the true size of u_j (see (13.2)) can be connected with an estimate for a certain arithmetic zeta-function on the critical line, and it puts the Lindelöf hypothesis for zeta-functions into a new perspective. For us (believers in the Riemann hypothesis, a fortiori, in the Lindelöf hypothesis) this connection insinuates that there is some chaos in the eigenstates on arithmetic surfaces (on the modular surface, at any rate).

13.2. Standard bounds.

Throughout we assume that $\{u_j(z)\}$ is an orthonormal system of cusp forms for the modular group which are eigenfunctions of all the Hecke operators. In this case (see Proposition 7.2)

$$(13.1) \qquad \sum_{0<t_j<T} |u_j(z)|^2 + \frac{1}{2\pi} \int_0^T |E(z, \frac{1}{2}+it)|^2\, dt \ll T^2 + Ty.$$

This shows that the u_j's are bounded on average. In the other direction we have proved recently that the $u_j(z)$ are not bounded individually; more precisely, at special arithmetic points (the complex multiplication points) one has (see [Iw-Sa])

$$|u_j(z)| \gg (\log \log \lambda_j)^{1/2}$$

for infinitely many j. Recently V. A. Bykovsky (1995) improved this lower bound to $\log \lambda_j$, for special points. Concerning an upper bound, we conjecture that

$$(13.2) \qquad |u_j(z)| \ll \lambda_j^\varepsilon$$

for any $\varepsilon > 0$ and $z \in \mathbb{H}$; the implied constant depends on ε and z. This estimate, if true, is very deep, since it has the Lindelöf hypothesis for certain L-functions as a consequence. From (13.1) we can only deduce (13.2) with $\varepsilon = 1/2$, while we already know that it holds with $\varepsilon = 1/4$. The implied constant in (13.2) must depend on z. Indeed, for $z = \frac{1}{4} + \frac{it_j}{2\pi} + o(1)$ we have

$$\lambda_j^{\frac{1}{12}-\varepsilon} \ll u_j(z) \ll \lambda_j^{\frac{1}{12}+\varepsilon}.$$

A sharper estimate will follow if we restrict the spectral averaging to a short interval. For this purpose we need the complete spectral decomposition of an automorphic kernel

(13.3)
$$K(z,z) = \sum_{\gamma \in \Gamma} k(u(\gamma z, z))$$
$$= \sum_{j} h(t_j) |u_j(z)|^2 + \frac{1}{2\pi} \int_{-\infty}^{+\infty} h(t) \left| E\left(z, \frac{1}{2} + it\right) \right|^2 dt.$$

We choose

(13.4)
$$h(t) = 4\pi^2 \frac{\cosh(\pi t/2) \cosh \pi T/2}{\cosh \pi t + \cosh \pi T},$$

so that $h(t) > 0$ everywhere and $h(t) \asymp 1$ if $t = T + O(1)$. The Fourier transform of $h(t)$ is equal to

(13.4)
$$g(x) = 2\pi \frac{\cos xT}{\cosh x}.$$

Hence one can show that the Selberg/Harish-Chandra transform satisfies

(13.5)
$$k(0) = T + O(1),$$

(13.6)
$$k(u) \ll T^{1/2} u^{-1/4} (1 + u)^{-5/4}.$$

Moreover, by Corollary 2.12 one obtains

(13.7)
$$\#\{\gamma \in \Gamma : \ 4u(\gamma z, z) + 2 \leq X\} \ll X,$$

for any $X \geq 2$, the implied constant depending on z. Using these estimates, we evaluate $K(z,z)$ as follows:

$$K(z,z) = \nu\, k(0) + \sum_{\substack{\gamma \in \Gamma \\ \gamma z \neq z}} k(u(\gamma z, z)) = \nu\, T + O(T^{1/2}),$$

where $\nu = 1, 2, 3$ is the order of the stability group of z. Combining this with the spectral expansion, we get

$$\sum_{j} h(t_j) |u_j(z)|^2 + \frac{1}{4\pi} \int_{-\infty}^{+\infty} h(t) \left| E\left(z, \frac{1}{2} + it\right) \right|^2 dt = \nu\, T + O(T^{1/2}).$$

Hence,

(13.8)
$$\sum_{T < t_j < T+1} |u_j(z)|^2 \ll T.$$

Recall that the segment $(T, T+1)$ contains $\asymp T$ points t_j, counted with multiplicity. This shows that the $u_j(z)$ are bounded on average over segments of constant length. Ignoring all but one term, we recover the standard bound (somewhat differently than the derivation of (8.3'))

$$(13.9) \qquad\qquad |u_j(z)| \ll \lambda_j^{1/4} \, .$$

This bound is as sharp as one can get by playing with test functions alone, save for a factor $(\log \lambda_j)^{-1}$.

The same bound can be established for eigenstates on any compact Riemann surface X. In such generality the exponent $1/4$ is best possible, as the example of the sphere $X = S^2$ shows. If, however, X is negatively curved, then a stronger bound is probably true.

13.3. Applying the Hecke operator.

In order to get further improvement we use the Hecke operator T_n as in the proof of Theorem 12.4. From the spectral decomposition we get

$$n^{-1/2} K_n(z, z) = \sum_j h(t_j) \, \lambda_j(n) \, |u_j(z)|^2$$
$$+ \frac{1}{4\pi} \int_{-\infty}^{+\infty} h(t) \, \eta_t(n) \, |E(z, \tfrac{1}{2} + it)|^2 \, dt \, ,$$

and on the geometric side we get

$$K_n(z, z) = \sum_{\gamma \in \Gamma_n} k(u(\gamma z, z)) = \nu_n \, k(0) + \sum_{\substack{\gamma \in \Gamma_n \\ \gamma z \neq z}} k(u(\gamma z, z)),$$

where

$$\nu_n = \#\{\gamma \in \Gamma_n : \ \gamma z = z\} \, .$$

We are looking for estimates which are uniform in n. To simplify, we consider only the special point $z = i$. In this case

$$\nu_n = \#\{a^2 + b^2 = n\} = r(n) \, .$$

For $\gamma = \left(\begin{smallmatrix} a & b \\ c & d \end{smallmatrix}\right) \in \Gamma_n$ with $\gamma i \neq i$ we have

$$(13.10) \qquad \begin{aligned} 4 \, n \, u(\gamma i, i) &= a^2 + b^2 + c^2 + d^2 - 2n \\ &= (a - d)^2 + (b + c)^2 \geq 1 \, . \end{aligned}$$

Lemma 13.1. *For $z = i$ and $X \geq 2$ we have*

$$\#\{\gamma \in \Gamma_n : \ 4 u(\gamma z, z) + 2 \leq X\} \ll \tau(n) \, n \, X \log X \, .$$

Proof. We are counting the number of solutions to $ad - bc = n$ within the ball $a^2 + b^2 + c^2 + d^2 \leq nX$. Use your fingers.

From (13.10), Lemma 13.1, (13.5) and (13.6) we infer an estimate for the geometric side of $K_n(z, z)$; then, combining this estimate with the spectral decomposition, we conclude the following:

Proposition 13.2. *For $z = i$ and $n \geq 1$ we have*

$$(13.11) \qquad \sum_j h(t_j)\, \lambda_j(n)\, |u_j(z)|^2 = \frac{r(n)}{\sqrt{n}}\, T + O(T^{1/2} n^{3/4+\varepsilon}),$$

the implied constant depending on ε alone.

We have dropped the contribution from the Eisenstein series, because it is easily absorbed by the error term (even more easily, one could drop it later, after the positivity is restored).

Since $1 \leq r(n) \leq 4\,\tau(n) \ll n^\varepsilon$, the formula (13.11) shows that as n gets large there exists a considerable cancellation of spectral terms due to the variation in sign of the Hecke eigenvalue $\lambda_j(n)$. This variation is the key to improving (13.9). Unfortunately, for the same reason, we cannot drop all but one term to obtain directly a good bound for the individual cusp form.

13.4. Constructing an amplifier.

We shall overcome the lack of positivity on the spectral side of (13.11) by means of an amplifier. First, using the multiplication rule for the eigenvalues

$$\lambda_j(m)\, \lambda_j(n) = \sum_{d|(m,n)} \lambda_j\!\left(\frac{mn}{d^2}\right),$$

we generalize (13.11) as follows:

$$\sum_j h(t_j)\, \lambda_j(m)\, \lambda_j(n)\, |u_j(z)|^2 = \frac{r(m,n)}{\sqrt{mn}}\, T + O(T^{1/2}(mn)^{3/4+\varepsilon}),$$

where

$$r(m,n) = \sum_{d|(m,n)} d\, r\!\left(\frac{mn}{d^2}\right).$$

Now multiply this throughout by any complex numbers a_m, $\bar{a}_n$ and sum over $m, n \leq N$, getting

$$\sum_j h(t_j) \left| \sum_{n \leq N} a_n \lambda_j(n) \right|^2 |u_j(z)|^2$$

$$= T \sum_{m,n \leq N} a_m \bar{a}_n \frac{r(m,n)}{\sqrt{mn}} + O(T^{1/2} N^{3/2+\varepsilon} \|\mathbf{a}\|_1^2),$$

where $\|\mathbf{a}\|_1$ denotes the ℓ_1-norm. Next apply $r(m,n) \ll (m,n)\,(mn)^\varepsilon$ and Cauchy's inequality to the leading term, getting

$$\sum_j h(t_j) \Big| \sum_{n \leq N} a_n\,\lambda_j(n) \Big|^2 |u_j(z)|^2 \ll \big(T\,\|\mathbf{a}\|^2 + T^{1/2}N^{3/2}\|\mathbf{a}\|_1^2\big)\,N^\varepsilon\,.$$

On the left side the terms are non-negative, so we can drop all but one, getting

$$(13.12) \qquad L_j^2|u_j(z)|^2 \ll \big(t_j\|\mathbf{a}\|^2 + t_j^{1/2}N^{3/2}\|\mathbf{a}\|_1^2\big)\,N^\varepsilon,$$

where

$$L_j = \sum_{n \leq N} a_n\,\lambda_j(n)\,.$$

This linear form L_j serves to amplify the contribution of the selected eigenform u_j. We would like to make L_j large; therefore, $a_n = \lambda_j(n)$ seems to be the obvious choice (no cancellation occurs). One expects that

$$(13.13) \qquad t_j^{-\varepsilon}N \ll \sum_{n \leq N} \lambda_j(n)^2 \ll t_j^\varepsilon N$$

uniformly in j and N, with the implied constant depending on ε alone. The upper bound has been proved in Theorem 8.3, but the lower bound is still a conjecture. If we accept this conjecture, then $L_j \gg t_j^{-\varepsilon}N$. We also have $\|\mathbf{a}\|^2 = L_j$ and $\|\mathbf{a}\|_1^2 \leq NL_j$, whence

$$|u_j(z)|^2 \ll \big(t_jN^{-1} + t_j^{1/2}N^{3/2}\big)\big(Nt_j\big)^\varepsilon$$

by (13.12). Choosing $N = t_j^{1/5}$, we conclude that

$$(13.14) \qquad |u_j(z)| \ll \lambda_j^{1/5+\varepsilon}$$

at the special point $z = i$. This is a conditional result, subject to the conjectured lower bound of (13.13).

Without any conjecture we have still a good choice, namely

$$a_n = \begin{cases} \lambda_j(p) & \text{if } n = p \leq \sqrt{N}\,, \\ -1 & \text{if } n = p^2 \leq N\,, \end{cases}$$

and we put $a_n = 0$ otherwise. Since $\lambda_j(p)^2 - \lambda_j(p^2) = 1$, the above choice yields

$$L_j = \sum_{p \leq \sqrt{N}} \big(\lambda_j(p)^2 - \lambda_j(p^2)\big) \sim \frac{N^{1/2}}{2\log N}\,.$$

Moreover, we have

$$\|\mathbf{a}\|^2 = \sum_{p \leq \sqrt{N}} \left(\lambda_j(p)^2 + 1 \right) \ll t_j^\varepsilon N^{1/2}$$

by the upper bound of (13.13), and

$$\|\mathbf{a}\|_1^2 = \sum_{p \leq \sqrt{N}} \left(|\lambda_j(p)| + 1 \right) \leq N^{1/4} \|\mathbf{a}\| .$$

Inserting these estimates into (13.12), we get

$$|u_j(z)|^2 \ll \left(t_j N^{-1/2} + t_j^{1/2} N^{3/2} \right) N^\varepsilon .$$

Choosing $N = t_j^{1/4}$, we obtain the following unconditional result:

Theorem 13.3. *For $z = i$ we have*

$$(13.15) \qquad\qquad |u_j(z)| \ll \lambda_j^{7/32+\varepsilon} ,$$

the implied constant depending only on ε.

Remarks. Using more refined estimates, one can get (13.15) with the exponent $5/24$ in place of $7/32$. Also the result holds true for any $z \in \mathbb{H}$, so it yields a bound for the $\mathfrak{L}^\infty$-norm:

$$(13.16) \qquad\qquad \|u_j\|_\infty \ll \lambda_j^{5/24+\varepsilon} .$$

The same estimate has been established for the eigenfunctions with respect to the quaternion group [Iw-Sa].

13.5. The ergodicity conjecture.

The next natural problem to consider after the conjecture (13.2) is that of equidistribution of $|u_j(z)|^2$ and $|E(z, 1/2 + it)|^2$. In particular, a basic question is what are the limits of the measures

$$(13.17) \qquad \begin{aligned} d\mu_j(z) &= |u_j(z)|^2 \, d\mu(z), \\ d\nu_t(z) &= |E(z, 1/2 + it)|^2 \, d\mu(z). \end{aligned}$$

Here u_j is normalized so that $d\mu_j$ has total mass equal to 1. The total mass of $d\nu_t$ is infinity, since the Eisenstein series cannot be normalized. In the language of quantum mechanics $d\mu_j$ represents the probability density of a particle being in the state u_j. Z. Rudnick and P. Sarnak [Ru-Sa] suggest that (the unique ergodicity conjecture)

$$(13.18) \qquad\qquad d\mu_j \to d\mu, \qquad \text{as } j \to \infty.$$

This has been proved for subsequences of full density (the quantum ergodicity theorem) by A. I. Shnirelman, S. Zelditch and Y. Colin de Verdière independently in various contexts. About the continuous spectrum, P. Sarnak has established that

$$(13.19) \qquad\qquad \nu_t(A) \sim \frac{48}{\pi}\, \mu(A) \log t, \qquad \text{as } t \to +\infty,$$

for any Jordan measurable set $A \subset X = \Gamma \backslash \mathbb{H}$. He also gave evidence for the unique ergodicity conjecture, by relating it to the Lindelöf hypothesis in the theory of L-functions. By means of the Fourier coefficients of u_j, the unique ergodicity conjecture requires essentially that there is a cancellation in the sums

$$\sum_{1 \le n \le N} \bar\nu_j(n)\, \nu_j(n+h)$$

for any fixed $h \neq 0$ and $N \asymp \lambda_j^{1/2}$ (these are building blocks for shifted convolution L-functions). Therefore, the classical problem concerning bounds for such sums shows a new face. A usual heuristic about oscillatory sums led Sarnak to the stronger quantitative form of the conjecture, namely that

$$(13.20) \qquad \int_X f(z)\, d\mu_j(z) = \int_X f(z)\, d\mu(z) + O(\lambda_j^{-1/4+\varepsilon})$$

for any $f \in C_0^\infty(X)$. Then, together with W. Luo [Lu-Sa], he showed that the above approximation holds true on average with respect to the spectrum. More precisely,

Theorem 13.4 (Luo and Sarnak). *Let* $f \in C_0^\infty(X)$ *be orthogonal to constants, i.e.* $\int_X f\, d\mu = 0$. *Then*

$$\sum_{t_j \le T} \left| \int_X f\, d\mu_j \right|^2 \ll T^{1+\varepsilon},$$

where the implied constant depends on ε *and* f.

They also gave estimates for discrepancy, showing that the exponent $-1/4$ in (13.20) is best possible.

The arithmetic quantum chaos (see a stimulating article by P. Sarnak [Sa2]) is a new area, in which modern number theory interacts with physics more strongly than ever before. Recently T. Watson (thesis, 2002) established a formula which reduces the ergodicity conjecture to a subconvexity bound for an L-function on GL_6 (a factor of the triple Rankin-Selberg convolution).

Classical Analysis

In the lectures we have appealed to several facts from classical analysis. We give here a brief account of these facts.

A.1. Self-adjoint operators.

Let $T : \mathcal{H} \to \mathcal{H}$ be a linear operator in a Hilbert space $\mathcal{H}$. Denote by $D(T)$ and $R(T)$ the domain and the range of T. If $D(T)$ is dense in $\mathcal{H}$, then there exists the adjoint operator T^* defined uniquely (Riesz theorem) by

$$\langle Tf, g \rangle = \langle f, T^* g \rangle.$$

Moreover, $T_1 \subset T_2$ implies $T_2^* \subset T_1^*$, and $T \subset T^{**}$.

An operator T is said to be *symmetric* if $T \subset T^*$ and *self-adjoint* if $T = T^*$.

Lemma A.1. *The eigenvalues of a symmetric operator are real.*

Proof. If $Tf = \lambda f$ with $f \in \mathcal{H}$, $f \neq 0$ and $\lambda \in \mathbb{C}$, then

$$\lambda \langle f, f \rangle = \langle \lambda f, f \rangle = \langle Tf, f \rangle = \langle f, T^* f \rangle = \langle f, Tf \rangle = \bar{\lambda} \langle f, f \rangle,$$

whence $\lambda = \bar{\lambda}$, as claimed.

Lemma A.2. *The eigenfunctions for distinct eigenvalues of a symmetric operator are orthogonal.*

Proof. Let $Tf = \lambda f$ and $Tg = \eta g$ with $f, g \in \mathcal{H}$. Then we have

$$\lambda \langle f, g \rangle = \langle Tf, g \rangle = \langle f, Tg \rangle = \eta \langle f, g \rangle,$$

whence $\langle f, g \rangle = 0$ if $\lambda \neq \eta$, as claimed.

A symmetric operator T is said to be *non-negative* if

$$\langle Tf, f \rangle \geq 0, \qquad \text{for all} \quad f \in D(T).$$

Therefore, the eigenvalues of a symmetric, non-negative operator are non-negative

Theorem A.3 (Friedrichs). *A symmetric, non-negative operator admits a self-adjoint extension.*

Now let T be a self-adjoint operator in a Hilbert space $\mathcal{H}$. The operator

$$R_\lambda = (T - \lambda)^{-1}$$

is called the *resolvent*. The complex number λ is said to be a *regular point for T* if the resolvent R_λ is defined on the whole space $\mathcal{H}$ and is a bounded operator. The remaining complex numbers comprise the *spectrum of T*; namely,

$$\sigma(T) = \big\{ \lambda \in \mathbb{C} : \ \lambda \ \text{not regular} \big\}.$$

The spectrum $\sigma(T)$ is partitioned into the *point spectrum* and the *continuous spectrum* (not a disjoint partition, in general) as follows:

- λ belongs only to the point spectrum if and only if R_λ is defined on a set not dense in $\mathcal{H}$ and it is bounded.

- λ belongs only to the continuous spectrum if and only if R_λ is defined on a dense set in $\mathcal{H}$ and it is unbounded.

- λ belongs to both spectra if and only if R_λ is defined on a set not dense in $\mathcal{H}$ and it is unbounded.

Lemma A.4. *Let T be a self-adjoint operator in $\mathcal{H}$. All $\lambda \in \mathbb{C} \setminus \mathbb{R}$ are regular points of T. Moreover,*

$$(A.1) \qquad\qquad\qquad \|R_\lambda\| \leq |\operatorname{Im} \lambda|^{-1}.$$

Proof. For any $\lambda \in \mathbb{C}$ we have

$$\|(T - \lambda)f\|^2 = \|(R_\lambda)^{-1}f\|^2 = \|(T - \operatorname{Re}\lambda)f\|^2 + (\operatorname{Im}\lambda)^2\|f\|^2.$$

If $\lambda \notin \mathbb{R}$, this shows that the operator $T - \lambda$ has zero kernel, the resolvent is defined on $(T - \lambda)\,D(T)$ and it is bounded by $|\operatorname{Im}\lambda|^{-1}$. It remains to show that $(T - \lambda)\,D(T) = \mathcal{H}$. Suppose $g \in \mathcal{H}$ is orthogonal to $(T - \lambda)\,D(T)$, so

$$\langle (T - \lambda)f, g \rangle = 0, \qquad \text{for any } f \in D(T).$$

This gives $\langle Tf, g \rangle = \langle f, \bar{\lambda}g \rangle$. Since $T = T^*$, it follows that $g \in D(T^*) = D(T)$ and $\bar{\lambda}g = T^*g = Tg$, so $\bar{\lambda}$ would be an eigenvalue of T if $g \neq 0$. Hence $g = 0$, which shows that $(T - \lambda)D(T)$ is dense in $\mathcal{H}$. But $(T - \lambda)D(T)$ is closed; therefore, it is equal to $\mathcal{H}$.

Lemma A.5 (Hilbert formula). *The resolvent of a self-adjoint operator at regular points satisfies*

$$(A.2) \qquad R_\lambda - R_\gamma = (\lambda - \gamma)R_\lambda R_\gamma \ .$$

Proof. This follows by applying R_λ to the obvious identity

$$I - (T - \lambda)R_\gamma = \big((T - \gamma) - (T - \lambda)\big)R_\gamma = (\lambda - \gamma)R_\gamma \ .$$

Lemma A.6. *Let T be a self-adjoint operator on $\mathcal{H}$ and $f, g \in \mathcal{H}$. The map*

$$\lambda \mapsto \langle f, \overline{R_\lambda g} \rangle$$

is holomorphic in $\mathbb{C} \setminus \mathbb{R}$.

Proof. Using Hilbert's formula, we infer the following relation:

$$(\gamma - \lambda)^{-1}\big(\langle f, \overline{R_\gamma g} \rangle - \langle f, \overline{R_\lambda g} \rangle\big) = \langle f, \overline{R_\lambda R_\gamma g} \rangle$$
$$= \langle f, \overline{R_\lambda R_\lambda g} \rangle - \overline{(\gamma - \lambda)}\langle f, \overline{R_\lambda R_\lambda R_\gamma g} \rangle \ ,$$

where the last inner product is bounded by $|\mathrm{Im}\,\lambda|^{-2}|\mathrm{Im}\,\gamma|^{-1}\|f\|\,\|g\|$. Hence, it is plain that the limit exists as $\gamma \to \lambda$, and it is equal to

$$\frac{d}{d\lambda}\langle f, \overline{R_\lambda g} \rangle = \langle f, \overline{R_\lambda R_\lambda g} \rangle = \langle R_\lambda f, \overline{R_\lambda g} \rangle$$

because $R_\lambda^* = R_{\bar{\lambda}}$.

A.2. Matrix analysis.

We recall a few basic facts about complex matrices (finite dimension operators) $A = (a_{ij}) \in M_n(\mathbb{C})$. We denote

$$\overline{A} = (\overline{a_{ij}}), \qquad \text{complex conjugate,}$$
$$A^t = (a_{ji}), \qquad \text{transpose,}$$
$$A^* = \overline{A}^t, \qquad \text{adjoint matrix.}$$

A is *non-singular* if it has the inverse A^{-1}, or equivalently its determinant $|A| = \det(a_{ij})$ does not vanish. A matrix $U \in M_n(\mathbb{C})$ is *unitary* if $U^*U = I$. It has the properties

- $|\det U| = 1$,

- the columns of U form an orthonormal set of vectors,

- x and Ux have the same length, *i.e.* U is an isometry, and

- the eigenvalues of U are on the unit circle.

Two matrices $A, B \in M_n(\mathbb{C})$ *commute* if $AB = BA$. A matrix $A \in M_n(\mathbb{C})$ is called *normal* if it commutes with its adjoint A^*. Clearly a unitary matrix is normal.

Proposition A.7 (Simultaneous diagonalization). *Let $\mathcal{F} \subset M_n(\mathbb{C})$ be a commuting family of normal matrices. There exists a unitary matrix U such that*

$$U^{-1}AU = \begin{pmatrix} \lambda_1 & & 0 \\ & \ddots & \\ 0 & & \lambda_n \end{pmatrix}, \qquad \text{for every } A \in \mathcal{F}.$$

A matrix is *Hermitian* if $A = A^*$. This means that the matrix A is normal, and it has all eigenvalues real. As a special case of Proposition A.7 we get

Corollary A.8 (Spectral theorem for Hermitian matrices). *Let A be a Hermitian matrix. There exist a diagonal real matrix*

$$\Lambda = \begin{pmatrix} \lambda_1 & & 0 \\ & \ddots & \\ 0 & & \lambda_n \end{pmatrix}$$

and a unitary matrix U such that $A = U\Lambda U^{-1}$. Moreover, if A is real, then U can be chosen real.

A slight generalization of this result is the following.

Corollary A.9. *Let L be a symmetric operator in a Hilbert space $\mathcal{H}$ with eigenspaces of finite dimension and let Δ be another symmetric operator in $\mathcal{H}$ which commutes with L. Then there exists a maximal orthonormal system of eigenfunctions of L (not necessarily a complete system in $\mathcal{H}$) which are also eigenfunctions of Δ.*

Proof. Let $\mathcal{H}_\lambda$ be the eigenspace of L for the eigenvalue λ. Since Δ commutes with L, it maps $\mathcal{H}_\lambda$ into itself. The unitary diagonalization of the corresponding Hermitian matrix yields the desired system.

A.3. The Hilbert-Schmidt integral operators.

Let F be a domain in $\mathbb{R}^2$. An integral operator

$$(Lf)(z) = \int_F k(z, w)\, f(w)\, dw$$

with a kernel $k \in \mathfrak{L}^2(F \times F, dz\, dw)$ is called a *Hilbert-Schmidt type operator*. Clearly $L : \mathfrak{L}^2(F) \to \mathfrak{L}^2(F)$, and it is a bounded operator, since

$$\|L\|^2 \le \iint\limits_{F \times F} |k(z, w)|^2\, dz\, dw\,.$$

Moreover, if $k(z, w) = \overline{k(z, w)}$, then L is symmetric (not necessarily self-adjoint).

Theorem A.10 (Hilbert-Schmidt). *Let $L \ne 0$ be a symmetric integral operator of Hilbert-Schmidt type. Then*

- *L has pure discrete spectrum,*

- *the eigenspaces of L have finite dimension,*

- *the eigenvalues of L can accumulate only at zero,*

- *L has at least one eigenvalue, the largest one being*

$$\mu_0 = \sup_{f \ne 0} \frac{\|Lf\|}{\|f\|} = \|L\|,$$

  where the supremum is attained by an eigenfunction of L (the variational principle),

- *the range of L in $\mathfrak{L}^2(F)$ is spanned by eigenfunctions of L. Let $\{u_j\}_{j \ge 0}$ be any maximal orthonormal system of eigenfunctions of L in $\mathfrak{L}^2(F)$, i.e.*

$$\langle u_j, u_k \rangle = \delta_{jk}\,, \quad Lu_j = \mu_j u_j \quad with \quad |\mu_0| \ge |\mu_1| \ge \cdots\,.$$

  Then any f from the range of L has an absolutely and uniformly convergent series representation

(A.3)
$$f(z) = \sum_{j \ge 0} \langle f, u_j \rangle\, u_j(z)\,.$$

A.4. The Fredholm integral equations.

Suppose λ is a complex number, D is a domain, $K : D \times D \longrightarrow \mathbb{C}$ is a kernel function, and $f : D \longrightarrow \mathbb{C}$ is a given function. The *Fredholm equation* (of the second type) is

$$(A.4) \qquad\qquad g(x) - \lambda \int_D K(x,y)\, g(y)\, dy = f(x),$$

or, in operator notation, $(I - \lambda K)g = f$. If $f \equiv 0$, then the equation is called *homogeneous*. We seek solutions $g \in \mathfrak{L}^2(D)$; therefore, it is natural to assume that $f \in \mathfrak{L}^2(D)$ and $K \in \mathfrak{L}^2(D \times D)$. Denote

$$\|K\|^2 = \iint\limits_{D \times D} |K(x,y)|^2 \, dx\, dy < +\infty \,.$$

Clearly $K : \mathfrak{L}^2(D) \longrightarrow \mathfrak{L}^2(D)$ is a compact operator.

The parameter $\lambda \in \mathbb{C}$ is said to be a *characteristic number* of the kernel $K(x,y)$ if the homogeneous equation

$$(A.5) \qquad\qquad\qquad (I - \lambda K)g = 0$$

has a solution $g \in \mathfrak{L}^2(D)$, $g \neq 0$. This is possible only if $\lambda \neq 0$, and then λ^{-1} is just the eigenvalue of the operator K, whereas g is its eigenfunction.

For small $|\lambda|$ the method of successive approximations leads to the solution. We start with $g_0 = f$ and define by induction $g_p = \lambda K g_{p-1} + f$, *i.e.*

$$g_p = \sum_{j=0}^{p} \lambda^j K^j f \,.$$

The infinite series

$$(A.6) \qquad\qquad\qquad g = \sum_{j=0}^{\infty} \lambda^j K^j f$$

is called the *Neumann series.* The norm of g is majorized by the series

$$\sum_{j=0}^{\infty} |\lambda|^j \|K\|^j \|f\| \,,$$

which converges absolutely in the disc $|\lambda| < \|K\|^{-1}$. For λ in this disc the Neumann series yields an $\mathfrak{L}^2$-solution. This solution is unique up to a function vanishing almost everywhere, because the homogeneous equation (A.5) implies

$$\|g\| \leq |\lambda| \, \|K\| \, \|g\|,$$

whence $g = 0$ almost everywhere since $|\lambda| \, \|K\| < 1$. In other words, this shows that the inverse operator $(I - \lambda K)^{-1}$ exists and is bounded by

$$\|(I - \lambda K)^{-1}\| \leq (1 - |\lambda| \, \|K\|)^{-1}$$

in the disc $|\lambda| < \|K\|^{-1}$. The iterated operators K^j are given by the kernels

$$(A.7) \qquad K_j(x, y) = \int_D K(x, z) \, K_{j-1}(z, y) \, dz, \quad j = 2, 3, \dots,$$

with $K_1 = K$. Suppose that

$$A(x)^2 = \int_D |K(x, y)|^2 \, dy < +\infty,$$

$$B(y)^2 = \int_D |K(x, y)|^2 \, dx < +\infty.$$

Applying the Cauchy-Schwarz inequality, we show by induction that

$$(A.8) \qquad |K_j(x, y)| \leq A(x) \, B(y) \, \|K\|^{j-2}, \qquad j = 2, 3, \dots.$$

Hence, the series

$$(A.9) \qquad R_\lambda(x, y) = \sum_{j=1}^{\infty} \lambda^{j-1} K_j(x, y)$$

is majorized by

$$|K(x, y)| + |\lambda| \, A(x) \, B(y) \sum_{j=0}^{\infty} |\lambda|^j \|K\|^j,$$

so it gives a function $R_\lambda(x, y)$ in $\mathcal{L}^2(D \times D)$, which is holomorphic in λ in the disc $|\lambda| < \|K\|^{-1}$. One can integrate term by term, showing by (A.6) that

$$(A.10) \qquad g(x) = f(x) + \lambda \int_D R_\lambda(x, y) \, f(y) \, dy.$$

In operator notation this asserts that $(I - \lambda K)^{-1} = I + \lambda R$, where R is the integral operator whose kernel is $R_\lambda(x, y)$. This kernel is called the *resolvent* of K, and it satisfies the Fredholm equation (use (A.9))

$$(A.11) \qquad R_\lambda(x, y) = K(x, y) + \lambda \int_D K(x, z) \, R_\lambda(z, y) \, dz.$$

By the principle of analytic continuation it follows that the solution g given by (A.10) is unique not only for λ in the disc $|\lambda| < \|K\|^{-1}$ but also in a larger domain to which R_λ has analytic continuation, and it is in $\mathfrak{L}^2(D \times D)$.

A nice, explicit construction was given by Fredholm in the special case of D having finite volume and $K(x,y)$ being bounded on $D \times D$,

$$\text{vol}\, D = V < +\infty \,,$$
(A.12)
$$|K(x,y)| \le K < +\infty \,.$$

In this case we shall construct two entire functions $\mathcal{D}(\lambda) \not\equiv 0$ and $\mathcal{D}_\lambda(x,y)$ such that

(A.13)
$$R_\lambda(x,y) = \mathcal{D}(\lambda)^{-1} \mathcal{D}_\lambda(x,y)$$

for all $\lambda \in \mathbb{C}$ with $\mathcal{D}(\lambda) \neq 0$. Put

$$K \begin{pmatrix} \xi_1, \ldots, \xi_m \\ \eta_1, \ldots, \eta_m \end{pmatrix} = \det \left(K(\xi_i, \eta_j) \right) .$$

By Hadamard's inequality we have

$$|\det (a_{ij})|^2 \le \prod_j \left(\sum_i |a_{ij}|^2 \right),$$

so we get

(A.14)
$$\left| K \begin{pmatrix} \xi_1, \ldots, \xi_m \\ \eta_1, \ldots, \eta_m \end{pmatrix} \right| \le (\sqrt{m}\, K)^m \,.$$

Let us denote

$$C_m = \int \cdots \int K \begin{pmatrix} \xi_1, \ldots, \xi_m \\ \xi_1, \ldots, \xi_m \end{pmatrix} d\xi_1 \cdots d\xi_m \,,$$

$$C_m(x,y) = \int \cdots \int K \begin{pmatrix} x, \xi_1, \ldots, \xi_m \\ y, \xi_1, \ldots, \xi_m \end{pmatrix} d\xi_1 \cdots d\xi_m \,.$$

By (A.14) we get

$$|C_m| \le (\sqrt{m}\, K\, V)^m \,,$$
$$|C_m(x,y)| \le (\sqrt{m+1}\, K)^{m+1} V^m \,.$$

Hence the series

(A.15)
$$\mathcal{D}(\lambda) = 1 + \sum_1^\infty \frac{(-\lambda)^m}{m!} C_m$$

is majorized by (use the inequality $(\sqrt{m}\,x)^m < m!\,e^{2x^2}$)

$$1 + \sum_{1}^{\infty} \frac{(\sqrt{m}\,|\lambda|\,K\,V)^m}{m!} < 2\,e^{8(|\lambda|KV)^2},$$

showing that it converges absolutely in the whole complex λ-plane, and

(A.16)
$$\mathcal{D}(\lambda) \ll \exp(3\,|\lambda|\,K\,V)^2.$$

Therefore, $\mathcal{D}(\lambda)$ is an entire function of order 2. Note that $\mathcal{D}(0) = 1$, so $\mathcal{D}(\lambda)$ does not vanish for small $|\lambda|$. Similarly, the series

(A.17)
$$\mathcal{D}_\lambda(x,y) = \sum_{0}^{\infty} \frac{(-\lambda)^m}{m!}\,C_m(x,y),$$

where $C_0(x,y) = K(x,y)$, is majorized by

$$e\,K \sum_{0}^{\infty} \frac{(\sqrt{m+1}\,|\lambda|\,K\,V)^m}{m!} \ll K \exp(3\,|\lambda|\,K\,V)^2,$$

showing that $\mathcal{D}_\lambda(x,y)$ is an entire function of order 2 in λ.

Developing the determinant by the first row, we obtain

$C_m(x,y)$

$$= \int \cdots \int \left(K(x,y)\,K\begin{pmatrix} \xi_1,\ldots,\xi_m \\ \xi_1,\ldots,\xi_m \end{pmatrix} \right.$$
$$\left. + \sum_{\ell=1}^{m} (-1)^\ell K(x,\xi_\ell)\,K\begin{pmatrix} \xi_1,\xi_2,\ldots\ldots\ldots,\xi_m \\ y,\xi_1,\ldots,\hat{\xi}_\ell,\ldots,\xi_m \end{pmatrix} \right) d\xi_1 \cdots d\xi_m$$
$$= C_m K(x,y)$$
$$- \sum_{\ell=1}^{m} \int K(x,\xi_\ell) \int \cdots \int K\begin{pmatrix} \xi_\ell,\xi_1,\ldots,\hat{\xi}_\ell,\ldots,\xi_m \\ y,\xi_1,\ldots,\hat{\xi}_\ell,\ldots,\xi_m \end{pmatrix} d\xi_1 \cdots d\xi_m$$
$$= C_m K(x,y) - \sum_{\ell=1}^{m} \int K(x,\xi_\ell)\,C_{m-1}(\xi_\ell,y)\,d\xi_\ell .$$

The positions of the careted elements above are meant to be skipped. Hence $C_m(x,y)$ satisfies the recurrence integral formula

$$C_m(x,y) = C_m K(x,y) - m \int K(x,z)\,C_{m-1}(z,y)\,dz$$

for all $m = 1, 2, \ldots$. Adding, we find that $\mathcal{D}_\lambda(x, y)$ satisfies the Fredholm equation

(A.18) $$\mathcal{D}_\lambda(x, y) = \mathcal{D}(\lambda) K(x, y) + \lambda \int K(x, z) \mathcal{D}_\lambda(z, y) \, dz \,,$$

which is the same one as for the resolvent $R_\lambda(x, y)$ (see (A.11)). Since for small $|\lambda|$ the Fredholm equation has a unique solution, we conclude that (A.13) is true for all $\lambda \in \mathbb{C}$ by analytic continuation.

Finally, from (A.10) and (A.13) we conclude that for any λ with $\mathcal{D}(\lambda) \neq 0$ the unique solution to the Fredholm equation is given by

(A.19) $$g(x) = f(x) + \frac{\lambda}{\mathcal{D}(\lambda)} \int_D \mathcal{D}_\lambda(x, y) \, f(y) \, dy \,.$$

A.5. Green function of a differential equation.

Generally speaking, a Green function $G_\lambda(z, z')$ is the kernel of the resolvent $R_\lambda = (I - \lambda T)^{-1}$ for a suitable linear operator T, provided, of course, the resolvent is an integral operator; and this, indeed, is the case for great many operators, either integral or differential. Precise definitions of a Green function vary a bit, as specific situations call for additional properties to meet the uniqueness requirement. The Green function depends analytically on λ in a small initial domain from which its analytic continuation leads the way to the spectral resolution of T.

In Section A.4 we constructed the Green function for Fredholm's integral equation. Now we consider the ordinary differential equation

(A.20) $$Tg = f,$$

where $T : C^\infty(\mathbb{R}^+) \longrightarrow C^\infty(\mathbb{R}^+)$ is a second order differential operator given by

(A.21) $$Tg(y) = -g''(y) + p(y) \, g(y), \qquad p \in C^\infty(\mathbb{R}^+) \,.$$

The Green function of the equation (A.20) (or of the operator T) is a function $G : \mathbb{R}^+ \times \mathbb{R}^+ \longrightarrow \mathbb{C}$ such that

(A.22) $$T \int_0^{+\infty} G(y, y') \, g(y') \, dy' = g(y),$$

for all $g \in C^\infty(\mathbb{R}^+)$ with $g(0) = g(+\infty) = 0$. Therefore G yields an integral operator which is the right inverse to T on functions g as above. Of course,

G is not unique. We shall be looking for the Green function which satisfies the following conditions:

(A.23)
$$G(y, y') \text{ is continuous in } \mathbb{R}^+ \times \mathbb{R}^+$$
$$\text{and smooth everywhere except for the diagonal } y = y',$$

(A.24)
$$TG(y, y') = 0 \qquad \text{if } y \neq y'.$$

Suppose a function $G(y, y')$ satisfies the above conditions. Differentiating the identity

$$\int_0^{+\infty} G(y, y')\, g(y')\, dy' = \int_0^y G(y, y')\, g(y')\, dy' + \int_y^{+\infty} G(y, y')\, g(y')\, dy',$$

we obtain

$$\frac{\partial}{\partial y} \int_0^{+\infty} G(y, y')\, g(y')\, dy'$$
$$= \int_0^y \frac{\partial}{\partial y} G(y, y')\, g(y')\, dy' + \int_y^{+\infty} \frac{\partial}{\partial y} G(y, y')\, g(y')\, dy',$$

because the terms $G(y, y-0)\, g(y)$ and $-G(y, y+0)\, g(y)$ cancel out by continuity. Differentiating again, we get

$$\frac{\partial^2}{\partial y^2} \int_0^{+\infty} G(y, y')\, g(y')\, dy' = \int_0^{+\infty} \frac{\partial^2}{\partial y^2} G(y, y')\, g(y')\, dy'$$
$$+ \left(\frac{\partial}{\partial y} G(y, y-0) - \frac{\partial}{\partial y} G(y, y+0) \right) g(y).$$

This yields the following identity:

$$T \int_0^{+\infty} G(y, y')\, g(y')\, dy' = \left(\frac{\partial}{\partial y} G(y, y-0) - \frac{\partial}{\partial y} G(y, y+0) \right) g(y).$$

Hence, we conclude that the property (A.22) is equivalent to

(A.25)
$$\frac{\partial}{\partial y} G(y, y-0) - \frac{\partial}{\partial y} G(y, y+0) = 1.$$

To construct a Green function satisfying the conditions (A.23) and (A.24) we take two linearly independent solutions to the homogeneous equation, say $I(y)$ and $K(y)$,

$$T\, I(y) = -I''(y) + p(y)\, I(y) = 0,$$
$$T\, K(y) = -K''(y) + p(y)\, K(y) = 0$$

and seek a solution of the type

$$G(y, y') = \begin{cases} I(y)\, A(y') & \text{if } y > y', \\ K(y)\, B(y') & \text{if } y < y', \end{cases}$$

where $A(y')$, $B(y')$ are to be determined. Then (A.24) is automatically satisfied. The continuity condition (A.23) and the jump condition (A.25) yield the linear system for the unknown functions

$$I(y)\, A(y) - K(y)\, B(y) = 0\,,$$
$$-I'(y)\, A(y) + K'(y)\, B(y) = 1\,.$$

The determinant of this system is the Wronskian $W = IK' - I'K$, which is constant because $W' = IK'' - I''K = 0$. Since I, K are linearly independent, $W \neq 0$. Hence, we have unique solutions $A(y) = W^{-1}K(y)$ and $B(y) = W^{-1}I(y)$ which give

$$G(y, y') = \begin{cases} W^{-1}\, I(y)\, K(y') & \text{if } y \geq y', \\ W^{-1}\, K(y)\, I(y') & \text{if } y \leq y'. \end{cases}$$

Special Functions

Special functions have been created gradually by the demand of scientists and engineers concerned with real computations. Various types of special functions are associated with specific problems of mechanics and physics. Due to the deep intuition of mathematicians the modern approach to special functions is unified beautifully through the language of group representation theory. Today special functions live on suitable symmetric spaces. Therefore, it is not surprising that some special functions are encountered in the theory of automorphic forms. Here we give excerpts from several sources of what is needed about special functions for these lectures. No proofs are supplied, so the reader still has to penetrate the jungle of relevant literature for completeness. Recommended books are [Gr-Ry], [Le], [Vi], [Wa].

B.1. The gamma function.

The gamma function of Euler is defined in $\operatorname{Re} s > 0$ as the Mellin transform of the exponential function:

$$(\text{B.1}) \qquad \Gamma(s) = \int_0^{+\infty} e^{-x}\, x^{s-1}\, dx\,.$$

It has a meromorphic continuation over the whole complex plane, given by

$$(\text{B.2}) \qquad \Gamma(s) = \int_1^{+\infty} e^{-x}\, x^{s-1}\, dx + \sum_{n=0}^{\infty} \frac{(-1)^n}{n!}\,(s+n)^{-1}\,.$$

Thus $\Gamma(s)$ has simple poles at non-positive integers. We have the Weierstrass product

(B.3)
$$s\,\Gamma(s) = \prod_{n=1}^{\infty} \left(1 + \frac{s}{n}\right)^{-1} \left(1 + \frac{1}{n}\right)^{s}$$
$$= e^{-\gamma s} \prod_{n=1}^{\infty} \left(1 + \frac{s}{n}\right)^{-1} e^{s/n}$$

where $\gamma = .5772156649\ldots$ is the Euler constant. Hence, $\Gamma(s)$ does not vanish anywhere, and so $\Gamma(s)^{-1}$ is an entire function of order 1. Also we have

Recursion formula:

(B.4)
$$s\,\Gamma(s) = \Gamma(s+1)\,.$$

Functional equation:

(B.5)
$$\Gamma(s)\,\Gamma(1-s) = \pi\,(\sin \pi s)^{-1}\,.$$

Duplication formula:

(B.6)
$$\Gamma(s)\,\Gamma(s + \frac{1}{2}) = \pi^{1/2}\,2^{1-2s}\,\Gamma(2s)\,.$$

Stirling's asymptotic formula

(B.7)
$$\Gamma(s) = \left(\frac{2\pi}{s}\right)^{1/2} \left(\frac{s}{e}\right)^{s} (1 + O(|s|^{-1}))$$

is valid in the angle $|\arg s| < \pi - \varepsilon$ with the implied constant depending on ε. Hence,

(B.8)
$$\Gamma(\sigma + it) = (2\pi)^{1/2}\,t^{\sigma-1/2}e^{-\pi t/2}\left(\frac{t}{e}\right)^{it}(1 + O(t^{-1}))$$

if $t > 0$, with the implied constant depending on σ.

The psi function is defined by

(B.9)
$$\psi(s) = \frac{\Gamma'}{\Gamma}(s) = -\gamma - \sum_{n=0}^{\infty} \left(\frac{1}{n+s} - \frac{1}{n+1}\right)\,.$$

It satisfies

(B.10)
$$\psi(s) = \frac{1}{m} \sum_{0 \le k < m} \psi\left(\frac{s+k}{m}\right) + \log m,$$

where m is any positive integer. We have the approximation

$$(B.11) \qquad \psi(s) = \log s - (2\,s)^{-1} + O(|s|^{-2})$$

uniformly in the angle $|\arg s| < \pi - \varepsilon$.

For $\operatorname{Re} u > 0$, $\operatorname{Re} v > 0$, $\operatorname{Re} s > 1/2$ we have

$$(B.12) \qquad \int_0^1 x^{u-1}(1-x)^{v-1}\,dx = \frac{\Gamma(u)\,\Gamma(v)}{\Gamma(u+v)}\,,$$

$$(B.13) \qquad \int_{-\infty}^{+\infty} (x^2+1)^{-s}\,dx = \pi^{1/2}\,\frac{\Gamma(s-1/2)}{\Gamma(s)}\,.$$

B.2. The hypergeometric functions.

A hypergeometric function is a solution of the differential equation

$$(B.14) \qquad z(1-z)\,F'' - ((\alpha+\beta+1)z - \gamma)\,F' - \alpha\beta\,F = 0,$$

where α, β, γ are complex numbers. If γ is not a non-positive integer, then one of the two linearly independent solutions is given by the Gauss hypergeometric series

$$(B.15) \qquad F(\alpha, \beta; \gamma; z) = \sum_{k=0}^{\infty} \frac{(\alpha)_k\,(\beta)_k}{(\gamma)_k\,k!}\,z^k$$

with coefficients given by $(s)_k = \Gamma(s+k)/\Gamma(s) = s \cdots (s+k-1)$. The power series converges absolutely in the unit disk $|z| < 1$. However, the hypergeometric function $F(\alpha, \beta; \gamma; z)$ has an analytic continuation over the plane $\mathbb{C}$ cut along $[1, +\infty)$. This is given by the integral representation

$$(B.16) \quad F(\alpha, \beta; \gamma; z) = \frac{\Gamma(\gamma)}{\Gamma(\beta)\,\Gamma(\gamma - \beta)} \int_0^1 t^{\beta-1}(1-t)^{\gamma-\beta-1}(1-tz)^{-\alpha}\,dt$$

when $\operatorname{Re} \gamma > \operatorname{Re} \beta > 0$, and by various recurrence formulas in the remaining cases. For example, the following formulas will do the job:

$$\alpha\,F(\alpha+1) - \beta\,F(\beta+1) = (\alpha-\beta)\,F\,,$$

$$\alpha\,F(\alpha+1) - (\gamma-1)\,F(\gamma-1) = (\alpha-\beta+1)\,F\,,$$

where only the shifted arguments are displayed; the other ones remain unchanged.

The hypergeometric function satisfies many transformations rules. Here are two examples:

$$(B.17) \qquad F(\alpha, \beta; \gamma; z) = (1-z)^{-\alpha} F\left(\alpha, \beta; \gamma; \frac{z}{1-z}\right),$$

$$(B.18) \quad F\left(2\alpha, 2\beta; \alpha + \beta + \frac{1}{2}; z\right) = F\left(\alpha, \beta; \alpha + \beta + \frac{1}{2}; 4z(1-z)\right).$$

These formulas are meaningful for $|z| < 1/2$. In extended ranges the values of $F(\alpha, \beta; \gamma, z)$ must be interpreted as those obtained by a suitable analytic continuation.

If either α or β but not γ is a non-positive integer, then the hypergeometric series terminates at a finite place; thus, it is a polynomial. In particular, we obtain

$$F\left(n+1, -n; 1; \frac{1-z}{2}\right) = P_n(z),$$

$$F\left(n, -n; \frac{1}{2}; \frac{1-z}{2}\right) = T_n(z),$$

the Legendre and Chebyshev polynomials, respectively.

For $|z| < 1$, we have

$$(B.19) \qquad F\left(\alpha, \alpha + \frac{1}{2}; 2\alpha + 1; 1 - z^2\right) = 4^{\alpha}(1+z)^{-2\alpha}.$$

If $\operatorname{Re}\gamma > 0$ and $\operatorname{Re}\gamma > \operatorname{Re}(\alpha + \beta)$, we obtain as $z \to 1-$ that

$$(B.20) \qquad F(\alpha, \beta; \gamma; 1) = \frac{\Gamma(\gamma)\,\Gamma(\gamma - \beta - \alpha)}{\Gamma(\gamma - \alpha)\,\Gamma(\gamma - \beta)}.$$

B.3. The Legendre functions.

Let ν, m be complex numbers. A solution to the differential equation

$$(B.21) \qquad (1-z^2)\,P'' - 2z\,P' + (\nu(\nu+1) - m^2(1-z^2)^{-1})\,P = 0$$

is called a *spherical function*. This equation is encountered in boundary value problems for domains of radial symmetry, for which solutions are naturally written in polar coordinates. We are interested exclusively in spherical functions with m a non-negative integer, in which case they are named Legendre functions of order m. The theory of Legendre functions of a positive order can be reduced to that of zero order. Indeed, if $P_\nu(z)$ satisfies (B.21) with $m = 0$, then

$$(B.22) \qquad P_\nu^m(z) = (z^2 - 1)^{m/2} \frac{d^m}{dz^m} P_\nu(z)$$

satisfies (B.21) with parameter m.

One of the solutions to (B.21) with $m = 0$ is given by the hypergeometric function:

$$(B.23) \qquad P_\nu(z) = F\left(-\nu, \nu + 1; 1; \frac{1-z}{2}\right).$$

This function is defined and analytic in the plane $\mathbb{C}$ cut along $(-\infty, -1]$. Then the Legendre function of order m derived from (B.22) is also defined and analytic in the same domain.

There are plenty of relations between Legendre functions, such as

$$P_\nu(z) = P_{-\nu-1}(z),$$

$$(\nu + 1)\, P_{\nu+1}(z) = (2\nu + 1)\, z\, P_\nu(z) - \nu\, P_{\nu-1}(z),$$

$$(z^2 - 1)\, P_\nu'(z) = \nu\, z\, P_\nu(z) - \nu\, P_{\nu-1}(z).$$

And there are useful integral representations, such as

$$(B.24) \qquad P_\nu^m(z) = \frac{\Gamma(\nu + m + 1)}{\pi\,\Gamma(\nu + 1)} \int_0^\pi \left(z + \sqrt{z^2 - 1}\,\cos\alpha\right)^\nu \cos(m\alpha)\, d\alpha.$$

If $\nu = n$ is a non-negative integer, then $P_n(z)$ is a polynomial of degree n given by

$$(B.25) \qquad P_n(z) = \frac{1}{2^n n!} \frac{d^n}{dz^n} (z^2 - 1)^n.$$

The generating function of the Legendre polynomials is

$$(B.26) \qquad \sum_{n=0}^\infty P_n(z)\, x^n = (1 - 2zx + x^2)^{-1/2}.$$

The Legendre functions $P_n^m(z)$ vanish identically if $m > n$. Those with $0 \le m \le n$, also called *spherical harmonics*, give a complete system of eigenfunctions of the Laplacian on the sphere $S^2 = \{x^2 + y^2 + z^2 = 1\}$. The eigenvalues are $\lambda = n(n+1)$, and the λ-eigenspace has dimension $2n+1$. In the polar coordinates $(x, y, z) = (\sin\theta\cos\varphi, \sin\theta\sin\varphi, \cos\theta)$ the Laplacian is given by

$$\frac{\partial^2}{\partial\theta^2} + \frac{1}{\tan\theta}\frac{\partial}{\partial\theta} + \frac{1}{(\sin\theta)^2}\frac{\partial^2}{\partial\varphi^2},$$

and the functions

$$\left(\left(n + \frac{1}{2}\right)\frac{(n-m)!}{(n+m)!}\right)^{1/2} P_n^m(\cos\theta)\, e^{\pm im\varphi}$$

with $0 \leq m \leq n$ form a complete, orthonormal system in the eigenspace $\lambda = n(n + 1)$. Notice that these eigenfunctions are not bounded in the spectrum. The maximum attained for $m = 0$ at $\theta = 0$ is equal to $(n + 1/2)^{1/2} = (\lambda + 1/4)^{1/4}$.

B.4. The Bessel functions.

The Bessel functions are solutions to the differential equation

$$(B.27) \qquad\qquad z^2 f'' + z f' + (z^2 - \nu^2) f = 0,$$

where ν is a complex number. In automorphic theory this equation arises when searching for eigenfunctions of the Laplace operator on the hyperbolic plane by the method of separation of variables in rectangular coordinates.

Since the equation (B.27) is singular at $z = 0$, there is going to be a problem. This is resolved by cutting the complex z-plane along the segment $(-\infty, 0]$. Then (B.27) has two linearly independent solutions which are holomorphic in $z \in \mathbb{C} \setminus (-\infty, 0]$. One of these is given by the power series

$$(B.28) \qquad\qquad J_\nu(z) = \sum_{k=0}^{\infty} \frac{(-1)^k}{k! \, \Gamma(k + 1 + \nu)} \left(\frac{z}{2} \right)^{\nu + 2k},$$

which converges absolutely in the whole plane. As a function of the parameter ν, called *the order of* $J_\nu(z)$, it is an entire function.

Changing ν into $-\nu$ does not alter the equation (B.27), so $J_{-\nu}(z)$ is another solution. The solutions $J_\nu(z)$, $J_{-\nu}(z)$ are linearly independent if and only if the Wronskian $W(J_\nu(z), J_{-\nu}(z)) = -2(\pi z)^{-1} \sin \pi \nu$ does not vanish identically, *i.e.* for ν different from an integer.

If $\nu = n$ is an integer, we have the relation

$$(B.29) \qquad\qquad J_n(z) = (-1)^n J_{-n}(z).$$

To get a pair of linearly independent solutions which would be suitable for all ν, we choose the linear combination

$$(B.30) \qquad Y_\nu(z) = (\sin \pi \nu)^{-1}(J_\nu(z) \cos \pi \nu - J_{-\nu}(z)),$$

where for $\nu = n$ the ratio is defined by taking the limit. Then $J_\nu(z)$ and $Y_\nu(z)$ are always linearly independent, since the Wronskian is

$$W(J_\nu(z), Y_\nu(z)) = 2(\pi z)^{-1}.$$

Changing the variable $z \mapsto iz$ (rotation of angle $\pi/2$), we transform the differential equation (B.27) into

$$(B.31) \qquad z^2 f'' + z f' - (z^2 + \nu^2) f = 0 \,.$$

As before, we seek holomorphic solutions in the complex plane cut along the negative axis. One of these is given by the power series

$$(B.32) \qquad I_\nu(z) = \sum_{k=0}^{\infty} \frac{1}{k!\, \Gamma(k + 1 + \nu)} \left(\frac{z}{2}\right)^{\nu + 2k} .$$

$I_\nu(z)$ and $I_{-\nu}(z)$ are linearly independent if and only if ν is not an integer. If $\nu = n$ is an integer, we have the relation

$$(B.33) \qquad I_n(z) = I_{-n}(z) \,.$$

We set

$$(B.34) \qquad K_\nu(z) = \frac{\pi}{2} \, (\sin \pi \nu)^{-1} (I_{-\nu}(z) - I_\nu(z))$$

and obtain a pair $I_\nu(z)$, $K_\nu(z)$ of linearly independent solutions to (B.31) for all ν, since the Wronskian is

$$W(I_\nu(z), K_\nu(z)) = -z^{-1} \,.$$

If $\nu = n$ is an integer, the functions $J_n(z)$ and $I_n(z)$ are actually entire in z. For ν not an integer there is a discontinuity along the negative axis; namely, we have

$$J_\nu(-x + \varepsilon i) - J_\nu(-x - \varepsilon i) \sim 2i \sin(\pi \nu) \, J_\nu(x)$$

for $x > 0$ as ε tends to zero. The same discontinuity appears for $I_\nu(z)$.

Each of the functions J_ν, Y_ν, I_ν, K_ν is expressible in terms of others in some way.

The Bessel functions of different order are related by many recurrence formulas. For the J-function we have

$$J_{\nu-1}(z) + J_{\nu+1}(z) = 2\,\nu\, z^{-1} J_\nu(z) \,,$$

$$J_{\nu-1}(z) - J_{\nu+1}(z) = 2\, J_\nu'(z) \,,$$

$$(z^\nu J_\nu(z))' = z^\nu J_{\nu-1}(z) \,,$$

$$(z^{-\nu} J_\nu(z))' = -z^{-\nu} J_{\nu+1}(z) \,.$$

The same formulas hold for the Y-function. For the I-function we have

$$I_{\nu-1}(z) - I_{\nu+1}(z) = 2\nu\, z^{-1}\, I_\nu(z)\,,$$

$$I_{\nu-1}(z) + I_{\nu+1}(z) = 2\, I_\nu'(z)\,,$$

$$(z^\nu I_\nu(z))' = z^\nu I_{\nu-1}(z)\,,$$

$$(z^{-\nu} I_\nu(z))' = z^{-\nu} I_{\nu+1}(z)\,.$$

And for the K-function the above formulas hold with negative sign on the right-hand sides.

The Bessel functions of half order are elementary functions:

$$J_{1/2}(z) = \left(\frac{2}{\pi z}\right)^{1/2} \sin z\,, \qquad Y_{1/2}(z) = -\left(\frac{2}{\pi z}\right)^{1/2} \cos z\,,$$

$$I_{1/2}(z) = \left(\frac{2}{\pi z}\right)^{1/2} \sinh z\,, \qquad K_{1/2}(z) = \left(\frac{\pi}{2z}\right)^{1/2} e^{-z}\,.$$

Applying the recurrence formulas, one can find elementary expressions for Bessel functions of any order ν which is half of an odd integer.

The four functions $J_\nu(y)$, $Y_\nu(y)$, $I_\nu(y)$, $K_\nu(y)$ in the real positive variable y have distinct asymptotic behaviour. For $y < 1 + |\nu|^{1/2}$ one will get good approximations by the first terms in the power series. For $y > 1 + |\nu|^2$ we have

$$\begin{aligned} J_\nu(y) &= \left(\frac{2}{\pi y}\right)^{1/2} \left(\cos\left(y - \frac{\pi}{2}\nu - \frac{\pi}{4}\right) + O\left(\frac{1 + |\nu|^2}{y}\right)\right), \\ Y_\nu(y) &= \left(\frac{2}{\pi y}\right)^{1/2} \left(\sin\left(y - \frac{\pi}{2}\nu - \frac{\pi}{4}\right) + O\left(\frac{1 + |\nu|^2}{y}\right)\right), \end{aligned}$$

(B.35)

and

$$\begin{aligned} I_\nu(y) &= (2\pi y)^{-1/2} e^y \left(1 + O\left(\frac{1 + |\nu|^2}{y}\right)\right), \\ K_\nu(y) &= \left(\frac{\pi}{2y}\right)^{1/2} e^{-y} \left(1 + O\left(\frac{1 + |\nu|^2}{y}\right)\right), \end{aligned}$$

(B.36)

where the implied constant is absolute. We don't have clear asymptotics for Bessel functions in the transition range $y \asymp 1 + |\nu|$.

There are various integral representations and transforms, each of which is useful in specific situations. Here is a selection for the K-function:

$$K_\nu(z) = \pi^{1/2}\Gamma(\nu + \frac{1}{2})^{-1}\left(\frac{z}{2}\right)^\nu \int_1^{+\infty} (t^2 - 1)^{\nu-1/2}e^{-tz}\,dt$$

$$= \pi^{-1/2}\Gamma(\nu + \frac{1}{2})\left(\frac{z}{2}\right)^{-\nu} \int_0^{+\infty} (t^2 + 1)^{-\nu-1/2}\cos(tz)\,dt$$

$$= \left(\frac{\pi}{2z}\right)^{1/2}\Gamma(\nu + \frac{1}{2})^{-1}e^{-z} \int_0^{+\infty} e^{-t}\left(t\left(1 + \frac{t}{2z}\right)\right)^{\nu-1/2}\,dt$$

$$= \frac{1}{2}\int_0^{+\infty} \exp\left(-\frac{z}{2}\left(t + \frac{1}{t}\right)\right)t^{-\nu-1}\,dt$$

$$= \int_0^{+\infty} e^{-z\cosh t}\cosh(\nu t)\,dt,$$

where $\operatorname{Re} z > 0$ and $\operatorname{Re}\nu > -1/2$. The Mellin transforms are often expressed as a product of gamma functions. For example, we have

$$\int_0^{+\infty} K_\nu(x)\,x^{s-1}\,dx = 2^{s-2}\,\Gamma(\frac{s+\nu}{2})\,\Gamma(\frac{s-\nu}{2}),$$

$$\int_0^{+\infty} e^{-x}K_\nu(x)\,x^{s-1}\,dx = 2^{-s}\pi^{1/2}\,\Gamma(s + \frac{1}{2})^{-1}\Gamma(s + \nu)\,\Gamma(s - \nu),$$

if $\operatorname{Re} s > |\operatorname{Re}\nu|$, and

$$\int_0^{+\infty} K_\mu(x)\,K_\nu(x)\,x^{s-1}\,dx = 2^{s-3}\Gamma(s)^{-1}\prod\Gamma(\frac{s\pm\mu\pm\nu}{2}),$$

if $\operatorname{Re} s > |\operatorname{Re}\mu| + |\operatorname{Re}\nu|$.

Similar formulas hold for the Mellin transform involving J-functions. One of these at $s = 0$ yields

$$(\text{B.37}) \qquad \int_0^{+\infty} J_\mu(x)\,J_\nu(x)\,x^{-1}\,dx = \frac{2\sin\pi(\mu - \nu)/2}{\pi\,(\mu - \nu)\,(\mu + \nu)},$$

if $\operatorname{Re}(\mu + \nu) > 0$. This formula shows, among other things, that the J-functions which are distinct but have the same order modulo even integers are orthogonal with respect to the measure $x^{-1}\,dx$.

B.5. Inversion formulas.

There are many ways of representing a function as a series and an integral of Bessel functions. We shall present three types of expansions. A general theory of eigenfunction expansions associated with second order differential equations is given by E.C. Titchmarsh [Ti2].

The Hankel inversion. Suppose f is a continuous function with finite variation on $\mathbb{R}^+$ such that

$$(\text{B.38}) \qquad \int_0^{+\infty} |f(x)|\, x^{-1/2}\, dx < +\infty\,.$$

The Hankel transform of f of order ν with $\operatorname{Re}\nu > -1/2$ is defined by

$$(\text{B.39}) \qquad H_f(y) = \int_0^{+\infty} f(x)\, J_\nu(xy)\, dx$$

for $y > 0$. It satisfies the inversion formula

$$(\text{B.40}) \qquad f(x) = \int_0^{+\infty} H_f(y)\, J_\nu(xy)\, xy\, dy\,.$$

The Kontorovitch-Lebedev inversion. This inversion is concerned with continuous representations in the K-functions of imaginary order. Suppose $f(y)$ is a smooth function with finite variation on $\mathbb{R}^+$ such that

$$(\text{B.41}) \qquad \int_0^{+\infty} |f(y)|\, (y^{-1/2} + y^{-1}|\log y|)\, dy < +\infty\,.$$

Then the integral transform

$$(\text{B.42}) \qquad L_f(t) = \int_0^{+\infty} K_{it}(y)\, f(y)\, y^{-1}\, dy$$

satisfies the inversion formula

$$(\text{B.43}) \qquad f(y) = \int_{-\infty}^{+\infty} L_f(t)\, K_{it}(y)\, \pi^{-2} \sinh(\pi t)\, t\, dt\,.$$

The Neumann series. The Neumann series is concerned with discrete representations in the J-functions of integral order. Note that $J_\ell(x)$ for $\ell > 0$ is square integrable on $\mathbb{R}^+$ with respect to the measure $x^{-1}\, dx$. The Neumann coefficients of a function $f \in \mathcal{L}^2(\mathbb{R}^+, x^{-1}\, dx)$ are defined by

$$(\text{B.44}) \qquad N_f(\ell) = \int_0^{+\infty} f(x)\, J_\ell(x)\, x^{-1}\, dx\,.$$

These give the Neumann series

$$(\text{B.45}) \qquad f^0(x) = \sum_{0 < \ell \text{ odd}} 2\ell\, N_f(\ell)\, J_\ell(x)\,.$$

If $f(x)$ is smooth on $\mathbb{R}^+$ and such that

$$(B.46) \qquad f^{(j)}(x) \ll x(x+1)^{-4}, \qquad 0 \le j \le 2,$$

then the Neumann series converges absolutely and uniformly. By the orthogonality property

$$2\ell \int_0^{+\infty} J_\ell(x)\, J_m(x)\, x^{-1}\, dx = \delta_{\ell,m}$$

if $\ell \equiv m \equiv 1 \pmod 2$ (see (B.37)), one shows that $f^0(x)$ is the projection of $f(x)$ on the subspace of $\mathcal{L}^2(\mathbb{R}^+, x^{-1}\, dx)$ spanned by the $J_\ell(x)$ of odd order. This subspace is not dense in $\mathcal{L}^2(\mathbb{R}^+, x^{-1}\, dx)$, so $f^0(x)$ is not always equal to $f(x)$.

The Titchmarsh integral. The Titchmarsh integral is concerned with continuous representations in the J-functions of imaginary order. Put

$$(B.47) \qquad B_\nu(x) = (2 \sin \frac{\pi}{2} \nu)^{-1} (J_{-\nu}(x) - J_\nu(x)).$$

Note that $B_{2it}(x) \in \mathcal{L}^2(\mathbb{R}^+, x^{-1}\, dx)$ if $t > 0$ and $B_{2it}(x)$ is orthogonal to all $J_\ell(x)$ with $0 < \ell \equiv 1 \pmod 2$, which fact follows by (B.37). The Titchmarsh coefficients of a function $f \in \mathcal{L}^2(\mathbb{R}^+, x^{-1}\, dx)$ are defined by

$$(B.48) \qquad T_f(t) = \int_0^{+\infty} f(x)\, B_{2it}(x)\, x^{-1}\, dx.$$

Therefore $T_f(t)\, B_{2it}(x)$ is the projection of $f(x)$ onto $B_{2it}(x)$. If $f(x)$ satisfies (B.46), we define the continuous superposition of these projections by the integral

$$(B.49) \qquad f^\infty(x) = \int_0^{+\infty} T_f(t)\, B_{2it}(x)\, \tanh(\pi t)\, t\, dt.$$

In other words, it turns out that $\tanh(\pi t)\, t\, dt$ is the relevant spectral measure. The continuous packet $\{B_{2it}\}_{t>0}$ spans densely the orthogonal complement to the linear subspace of the discrete collection $\{J_1, J_3, J_5, \dots\}$ in $\mathcal{L}^2(\mathbb{R}^+, x^{-1}\, dx)$. More precisely, for $f(x)$ satisfying (B.46), the Sears-Titchmarsh inversion holds:

$$(B.50) \qquad f(x) = f^0(x) + f^\infty(x).$$

The above partition can be characterized nicely in terms of the Hankel transform of order zero. Indeed, by recurrence formulas for the J-function we obtain

$$\frac{2\nu}{xy} \frac{d}{du} J_\nu(ux) J_\nu(uy) = u\, J_{\nu-1}(ux)\, J_{\nu-1}(uy) - u\, J_{\nu+1}(ux)\, J_{\nu+1}(uy).$$

Integrating over $0 < u < 1$ and summing over $\nu = 1, 3, 5, \ldots$, we get

(B.51) $$\sum_{0 < \ell \text{ odd}} 2\ell \, J_\ell(x) \, J_\ell(y) = xy \int_0^1 u \, J_0(ux) \, J_0(uy) \, du \,.$$

Hence,

(B.52) $$f^0(x) = \int_0^1 ux \, J_0(ux) \, H_f(u) \, du,$$

where $H_f(u)$ is the Hankel transform of f, given by

(B.53) $$H_f(u) = \int_0^{+\infty} f(y) \, J_0(uy) \, dy \,.$$

By the Hankel inversion (B.40) and the Sears-Titchmarsh inversion (B.50) it follows that

(B.54) $$f^\infty(x) = \int_1^{+\infty} ux \, J_0(ux) \, H_f(u) \, du \,.$$

Therefore the projections $f^0(x)$, $f^\infty(x)$ are obtained by truncating the Hankel transform of f to the segments $0 \leq u < 1$ and $1 \leq u < +\infty$ respectively.

References

[At-Le] A. O. L. Atkin and J. Lehner, Hecke operators on $\Gamma_0(m)$. *Math. Ann.* **185** (1970), 134-160.

[Br1] R. W. Bruggeman, Fourier coefficients of cusp forms. *Invent. Math.* **45** (1978), 1-18.

[Br2] R. W. Bruggeman, Automorphic forms, in *Elementary and Analytic Theory of Numbers*, Banach Center Pub. **17**, PWN, Warsaw, 1985, 31-74.

[Bu-Du-Ho-Iw] D. Bump, W. Duke, J. Hoffstein and H. Iwaniec, An estimate for the Hecke eigenvalues of Maass forms. *Duke Math. J. Research Notices* **4** (1992), 75-81.

[Ch] F. Chamizo, *Temas de Teoría Analítica de los Números*. Doctoral Thesis, Universidad Autónoma de Madrid, 1994.

[Co-Pi] J. W. Cogdell and I. Piatetski-Shapiro, *The Arithmetic and Spectral Analysis of Poincaré Series*. Perspectives in Math. **13**, Academic Press, 1990.

[De-Iw] J.-M. Deshouillers and H. Iwaniec, Kloosterman sums and Fourier coefficients of cusp forms. *Invent. Math.* **70** (1982), 219-288.

[El] J. Elstrodt, Die Resolvente zum Eigenwertproblem der automorphen Formen in der hyperbolischen Ebene, I, II, III. *Math. Ann.* **203** (1973), 295-330; *Math. Z.* **132** (1973), 99-134; *Math. Ann.* **208** (1974), 99-132.

[Fa] J. D. Fay, Fourier coefficients of the resolvent for a Fuchsian group. *J. Reine Angew. Math.* **293/294** (1977), 143-203.

[Fi] J. Fischer, *An Approach to the Selberg Trace Formula via the Selberg Zeta-Function,* Springer Lecture Notes in Math. **1253** (1980).

[Foc] V. A. Fock, On the representation of an arbitrary function by an integral involving Legendre's functions with a complex index. *C. R. (Dokl.) Acad. Sci. URSS* **39** (1943), 253-256.

[For] L. R. Ford, *Automorphic Functions*. McGraw-Hill, New York, 1929.

[Fr-Kl] R. Fricke and F. Klein, *Vorlesungen über die Theorie der automorphen Funktionen I, II.* Teubner, Leipzig, 1897 and 1912.

[Go]   A. Good, *Local Analysis of Selberg's Trace Formula*. Springer Lecture Notes in Math. **1040**, 1983.

[Go-Sa] D. Goldfeld and P. Sarnak, Sums of Kloosterman sums. *Invent. Math.* **71** (1983), 243-250.

[Gr-Ry] I. S. Gradshteyn and I. M. Ryzhik, *Table of Integrals, Series and Products*. Academic Press, London, 1965.

[Ha-Ra] G. H. Hardy and S. Ramanujan, Asymptotic formulae in combinatory analysis. *Proc. London Math. Soc.* **17** (1918), 75-115.

[Ha-Ti] G. H. Hardy and E. C. Titchmarsh, Solutions of some integral equations considered by Bateman, Kapteyn, Littlewood, and Milne. *Proc. London Math. Soc.* **23** (1924), 1-26.

[He1]   D. A. Hejhal, *The Selberg Trace Formula for $PSL(2,\mathbb{R})$*. Springer Lecture Notes in Math. **548** (1976) and **1001** (1983).

[He2]   D. A. Hejhal, *Eigenvalues of the Laplacian for Hecke triangle groups*. Memoirs of the Amer. Math. Soc. **469** (1992).

[Ho-Lo] J. Hoffstein and P. Lockhart, Coefficients of Maass forms and the Siegel zero. *Ann. of Math.* **140** (1994), 161-176.

[Hu1]   M. N. Huxley, Scattering matrices for congruence subgroups, in *Modular forms*, Ellis Horwood Series of Halsted Press, New York, 1984, 141-156.

[Hu2]   M. N. Huxley, Introduction to Kloostermania, in *Elementary and Analytic Theory of Numbers*. Banach Center Publ. **17**, Warsaw, 1985, 2217-306.

[Hu3]   M. N. Huxley, Exponential sums and lattice points, II. *Proc. London Math. Soc.* **66** (1993), 279-301; Corrigenda, *ibid.* **68** (1994), 264.

[Iw1]   H. Iwaniec, Non-holomorphic modular forms and their applications, in *Modular Forms*, Ellis Horwood Series of Halsted Press, New York, 1984, 157-196.

[Iw2]   H. Iwaniec, Spectral theory of automorphic functions and recent developments in analytic number theory. *Proc. I.C.M.*, Berkeley, 1986, 444-456.

[Iw3]   H. Iwaniec, Prime geodesic theorem. *J. Reine Angew. Math.* **349** (1984), 136-159.

[Iw-Sa] H. Iwaniec and P. Sarnak, L^∞-norms of eigenfuntions of arithmetic surfaces. *Ann. of Math.* **141** (1995), 301-320.

[Kac]   M. Kac, Can one hear the shape of a drum? *Amer. Math. Monthly* **73** (1966), 1-23.

[Kat]   S. Katok, *Fuchsian Groups*. Univ. of Chicago Press, 1992.

[Ki-Sa] H. Kim and P. Sarnak, Functoriality for the exterior square of GL_2 (by Henry Kim); with the appendix, Refined estimates towards the Ramanujan and Selberg conjectures (by Henry Kim and Peter Sarnak), to appear in *J.Amer.Math.Soc.*

[Kl1]   H. D. Kloosterman, On the representation of numbers in the form $ax^2 + by^2 + cz^2 + dt^2$. *Acta Math.* **49** (1926), 407-464.

[Kl2]   H. D. Kloosterman, Asymptotische Formel für die Fourier-Koeffizienten ganzer Modulformen. *Abh. Math. Sem. Univ. Hamburg* **5** (1927), 338-352.

[Ko-Le] M. J. Kontorovitch and N. N. Lebedev, *J. Exper. Theor. Phys. USSR* **8** (1938), 1192-1206, and **9** (1939), 729-741.

[Ku]   N. V. Kuznetsov, Petersson's conjecture for cusp forms of weight zero and Linnik's conjecture. Sums of Kloosterman sums. *Math. USSR Sbornik* **29** (1981), 299-342.

[L]   E. Landau, Über die Klassenzahl imaginär-quadratischer Zahlkörper, *Gött. Nachr.* (1918), 285-295.

[La]   S. Lang, $SL_2(\mathbb{R})$. Addison-Wesley, Reading, MA, 1973.

[Le]   N. N. Lebedev, *Special Functions and their Applications*. Dover Publications, New York, 1972.

[Lu]   W. Luo, On the non-vanishing of Rankin-Selberg L-functions, *Duke Math. J.* **69** (1993), 411-427.

[Lu-Ru-Sa] W. Luo, Z. Rudnick and P. Sarnak, On Selberg's eigenvalue conjecture, *Geom. Funct. Anal.* **5** (1995), 384-401. See also; On the generalized Ramanujan conjecture for $GL(n)$, *Proc. Symp. Pure Math.* **66**, part 2 (1999), 301-310.

[Lu-Sa] W. Luo and P. Sarnak, Quantum ergodicity of eigenfunctions on $SL_2(\mathbb{Z})\backslash\mathbb{H}^2$. *IHES Publ. Math.* **81** (1995), 207-237.

[Ma]   H. Maass, Über eine neue Art von nichtanalytschen automorphen Funktionen und die Bestimmung Dirichletscher Reihen durch Funktionalgleichungen. *Math. Ann.* **121** (1949), 141-183.

[Me]   F. G. Mehler, Über eine mit den Kugel- und Cylinderfunctionen verwandte Funktion und ihre Anwendung in der Theorie der Elektricitätsvertheilung. *Math. Ann.* **18** (1881), 161-194.

[Mi-Wa] R. Miatello and N. R. Wallach, Kuznetsov formulas for rank one groups. *J. Funct. Anal.* **93** (1990), 171-207.

[Ni]   J. Nielsen, Über Gruppen linearer Transfromationen. *Mitt. Math. Ges. Hamburg* **8** (1940), 82-104.

[Pe1]   H. Petersson, Über die Entwicklungskoeffizienten der automorphen Formen. *Acta Math.* **58** (1932), 169-215.

[Pe2]   H. Petersson, Über einen einfachen Typus von Untergruppen der Modulgruppe. *Archiv der Math.* **4** (1953), 308-315.

[Ph-Ru] R. S. Phillips and Z. Rudnick, The circle problem in the hyperbolic plane, *J. Funct. Anal.* **121** (1994), 78-116.

[Ph-Sa] R. S. Phillips and P. Sarnak, On cusp forms for cofinite subgroups of $PSL(2,\mathbb{R})$. *Invent. Math.* **80** (1985), 339-364.

[Pr]   N. V. Proskurin, Summation formulas for general Kloosterman sums. *Zap. Nauchn. Sem. LOMI* **82** (1979), 103-135; English transl., *J. Soviet Math.* **18** (1982), 925-950.

[Rad]   H. Rademacher, A convergence series for the partition function $p(n)$. *Proc. Nat. Acad. Sci. U.S.A.* **23** (1937), 78-84.

[Ran]   R. A. Rankin, *Modular Forms of Negative Dimensions*. Dissertation. Clare College, Cambridge, 1940.

[Ro]   W. Roelcke, Über die Wellengleichung be Grenzkreisgruppen erster Art. *S.-B. Heidelberger Akad. Wiss. Math. Nat. Kl.* **1953/55**, Abh. 4, 159-267 (1956).

[Ru-Sa] Z. Rudnick and P. Sarnak, The behavior of eigenstates of arithmetic hyperbolic manifolds, *Comm. Math. Phys.* **161** (1991), 195-213.

[Sa1] P. Sarnak, Determinants of Laplacians. *Comm. Math. Phys.* **110** (1987), 113-120.

[Sa2] P. Sarnak, *Arithmetic Quantum Chaos.* R. A. Blyth Lectures, Univ. of Toronto, 1993.

[Sa3] P. Sarnak, *Some Applications of Modular Forms.* Cambridge Tracts in Math. **99**, Cambridge University Press, 1990.

[Sa4] P. Sarnak, Class numbers of indefinite binary quadratic forms. *J. Number Theory* **15** (1982), 229-247.

[Se-Ti] D. B. Sears and E. C. Titchmarsh, Some eigenfunction formulae, *Quart J. Math. Oxford Ser.* (2) **1** (1950), 165-175.

[Se1] A. Selberg, On the estimation of Fourier coefficients of modular forms. *Proc. Symp. Pure Math.* **7** (1965), 1-15.

[Se2] A. Selberg, *Collected Papers,* vol. I. Springer, 1989.

[Se3] A. Selberg, Über die Fourierkoeffizienten elliptischer Modulformen negativer Dimension. *C. R. Neuvième Congrès Math. Scandinaves (Helsingfors, 1938)*, Mercatorin Kirjapaino, Helsinki, 1939, 320-322.

[Sh] G. Shimura, *Introduction to the Arithmetic Theory of Automorphic Functions.* Princeton Univ. Press, 1971.

[Si1] C. L. Siegel, Bemerkung zu einem Satze von Jakob Nielsen. *Mat. Tidsskr. B* **1950**, 66-70.

[Si2] C. L. Siegel, Discontinuous groups. *Ann. of Math.* **44** (1943), 674-689.

[Te] A. Terras, *Fourier Analysis on Symmetric Spaces and Applications, I, II.* Springer-Verlag, 1985, 1988.

[Ti1] E. C. Titchmarsh, *Introduction to the Theory of Fourier Integrals.* 2nd ed., Clarendon Press, Oxford, 1948.

[Ti2] E. C. Titchmarsh, *Eigenfunction Expansions Associated with Second-Order Differential Equations.* Vols. I (rev. ed.), II, Clarendon Press, Oxford, 1962, 1958.

[Ve] A. B. Venkov, *Spectral Theory of Automorphic Functions and its Applications.* Math. and Its Applications (Soviet Series), **51**, Kluwer, Dordrecht, 1990.

[Vi] N. Ya. Vilenkin, *Special Functions and Theory of Group Representations.* Transl. Math. Monographs **22**, Amer. Math. Soc., 1968.

[Vig1] M.-F. Vignéras, L'équation fonctionelle de la fonction zéta de Selberg du groupe modulaire $PSL(2,\mathbb{Z})$. *Asterisque* **61** (1979), 235-249.

[Vig2] M.-F. Vignéras, Variétés riemanniennes isospectrales et non isométriques. *Ann. of Math.* **112** (1980), 21-32.

[Wa] G. N. Watson, *A Treatise on the Theory of Bessel Functions.* Cambridge University Press, 1962.

[We] A. Weil, On some exponential sums. *Proc. Nat. Acad. Sci. U.S.A.* **34** (1948), 204-207.

[Wi]   A. Winkler, Cusp forms and Hecke groups. *J. Reine Angew. Math.* **386** (1988), 181-204.

[Wo]   S. Wolpert, Disappearence of cusp forms in families. *Ann. of Math.* **139** (1994), 239-291.

[Zo]   P. Zograf, Fuchsian groups and small eigenvalues of the Laplace operator. *Zap. Nauchn. Sem. LOMI* **122** (1982), 24-29; English transl., *J. Soviet Math.* **26** (1984), 1618-1621.

Recommended new books:

[Br3]   R. W. Bruggeman, *Families of Automorphic Forms.* Birkhäuser Monographs in Mathematics **88**, Basel, 1994.

[Bu]   D. Bump, *Automorphic Forms and Representations.* Cambridge University Press, 1997.

[Mo]   Y. Motohashi, *Spectral Theory of the Riemann Zeta-Function.* Cambridge University Press, 1997.

Recommended survey articles:

[Iw4]   H. Iwaniec, Harmonic analysis in number theory, in *Prospects in Mathematics* (Hugo Rossi, editor), AMS, Providence, RI, 1999, 51-68.

[Iwa-Sar] H. Iwaniec and P. Sarnak, Perspectives on the analytic theory of *L*-functions, in *Visions in Mathematics Towards 2000*, Birkhäuser, Basel, 2000, 705-741.

Subject Index

Notation Index